Learning the Secret Language of Cats
A Vet's Translation

Learning the Secret Language of Cats
A Vet's Translation

[Signed: Catherine — Best Wishes, Carol Teed]

Dr. Carol Teed D.V.M.

Foreword by Dr. Liz O'Brien D.V.M., DABVP (Feline Practice)

First Published in Canada 2013 by Influence Publishing

© Copyright Carol Teed

All rights reserved. No part of this publication may be reproduced, stored in or introduced into a retrieval system, or transmitted, in any form, or by any means (electronic, mechanical, photocopying, recording or otherwise) without the prior written permission of the publisher. This book is sold subject to the condition that it shall not, by way of trade or otherwise, be lent, resold, hired out, or otherwise circulated without the publishers prior consent in any form of binding or cover other than that in which it is published and without a similar condition including this condition being imposed on the subsequent purchaser.

Book Cover Design: Greg Salisbury & Divino Mucciante
Typeset: Greg Salisbury
Photographs by: Divino Mucciante

DISCLAIMER: This book is based on the author's own experiences working with and caring for cats. The author's opinions may not match the experience or opinions of other veterinarians, or others working in the field. The information in this book is not intended in any way to replace other professional advice. Readers of this publication agree that neither Carol Teed, nor her publisher, will be held responsible or liable for damages that may be alleged as resulting directly or indirectly from the use of this publication. Neither the publisher nor the author can be held accountable for the information provided by, or actions resulting from, accessing any mentioned resources. Except where permission has been asked and granted, all cat names have been changed for their privacy.

This book is for the cats, and my cat-like partners, Michelle and Jessica, who learned the language of cats alongside me as we did our most precious work. And for Mady, one of my dearest clients who encouraged me to write this book.

A note to my readers:
It is a great pleasure to share my book with you. Writing it has been a labour of love and I know already that we share a bond because you are reading it.

*"If man could be crossed with the cat,
it would improve man but deteriorate the cat."*
Mark Twain

"A cat never lacks grace- it never lacks dignity. It displays a sense of self, that, somehow, humans have either forgotten or never known. All you have is what you are. All you are is what you are given. All you were given is the certainty of self."
Timothy Findley, Spadework

What others are saying about Learning the Secret Language of Cats:

"With chapter titles including 'Cats are Like Potato Chips-you can't have just one' and 'All cats are grey in the dark', you know this is not just another book about cats! This book is educational, passionate and entertaining. It embraces not only the history, behavior, health and welfare, and diseases of the cat, but also the personal life story of Dr. Teed, a veterinarian, wife, mother and lover of all cats. Dr. Teed takes us on a personal journey detailing her connection to cats from a young age, her pathway to a professional life as a veterinarian and specifically, to owning a cat-only practice - no dogs allowed! Along the way, we gain a glimpse into the personal life of Dr. Teed, her family of 2 legged and 4 legged housemates and the role cats have played at every step along the way."
Doreen Houston, DVM, DVSc, Diplomate ACVIM (Internal Medicine)

"I am blessed to have had the opportunity to read this beautiful book that ...reveals what so many of us struggle with....that the real work of life is to open ourselves to living in our spiritual heart center, just like Carol Teed has experienced. This book provides insight through the eyes of a Cat and the wisdom of love... I began to read and just could not stop, it captured me. Carol has her heart wide open to feel every move the feline makes with the spark of yoga in her life. You feel her grounded passion, the energy of the Feline Guru of divine creativity who shares the sacred journey of life,..... and the understanding that it is love and action that can transform this planet from one of fear and separation into one united by truth, love, and God."
Lisa Devi Kirn Scandolari, International Yoga Teacher, Founder of Kundalini in Niagara

"Dr. Teed has unveiled the loving spirit of the feline. I thank her for sharing her extensive knowledge and experience, and her beautiful soul with the world."
Deborah-Marie Forrester, Reiki Practitioner

"Cats - who knew; I've always considered myself to be more of a dog person than a cat person, but after reading Dr. Teed's impassioned interpretation of the secret language of cats I have garnered a new appreciation for these mysterious creatures, and the beauty and wisdom they possess. This book is a must read for both dog lovers and cat lovers, it will bring a tear to your eye, a smile to your face, and give you a much better understanding of the inner and outer beauty of cats."
Randy Valpy, Pet Advocate and Pet Insurance Expert

"Definitely not just another book about cats, Dr. Teed has rewarded us with an eloquent tale of her experiences as a feline practitioner and long time "cat whisperer." Her compassion, insight and humour all culminate in a truly delightful read for anyone who loves cats, for those that own a cat but think they really don't like cats, and to all animal lovers. The book asks us to reflect on ourselves and shows us how we can benefit from our interactions with the feline species. This very well written tale made me laugh, cry, and appreciate just how much the beautiful cat can teach us at every stage of life."
Dr. Patti Murphy, Doctor of Veterinary Medicine

"...An intriguing exploration of the author's lifelong love of cats, and what she has learned from them. This book will be of interest to those who care about all things 'cat'."
Dr. Alice Crook, DVM, Adjunct Professor Atlantic Veterinary College, U.P.E.I., Coordinator of Sir James Dunn Animal Welfare Centre.

Acknowledgements

First, I want to thank the cats for inspiring me to write in the first place, and to find my words outside of the confines of a medical record.

In the process of writing, however, there are many humans I need to thank for helping me along the way. A big thank you to my publisher Julie Salisbury, and to my editors Samantha Michaels, Gulnar Patel, and Amy O'Hara, for their kindness and extreme patience through this process, for getting me organized, and for keeping me on track. To my husband Dennis, and children, Madeleine, Josie, David, and Kate for putting up with my absence while at the keyboard and for letting me talk about them from time to time. Thanks to my cat-like workmates Michelle Burgoyne and Jessica Soul for helping to jog my memory of all the wonderful cats we saw in the clinic and for cheering me on. And also to my many wonderful clients and friends who shared their cat stories and experiences with me over the years. A huge purring thanks to Dr. Patti Murphy, my very first reader, for her amazing support and belief in this book. Thanks also to Sue Lymburner for allowing me to talk about it endlessly and her good wishes over these many months, and also for allowing me to photograph Cosmo in a paper bag. Thank you to Divino Mucciante for your exceptional talent with a camera and for sharing your photos here in this book. Wow. Thanks to Dr. Alice Crook, Dr. Doreen Houston, Randy Valpy, Kathy Stinson, Heather Skoll, Deborah Forrester, and Lisa Devi Kirn Scandolari for reading sections of my book and providing meaningful feedback and suggestions. Many thanks also to Dr. Liz O'Brien for her wonderful insights and suggestions and for her very valuable contribution of writing the Foreword for this book. And finally thank you to Mady Fitzgibbon for putting this book idea into my head so many years ago. I did it. Now it is done. And maybe there will be another one someday.

Contents

Foreword
Prologue: Letting the Cat Out of the Bag

Chapter 1: The Purrfect Career for Me 1

Chapter 2: Cats are Like Potato Chips—You Can't Have Just One 9

Chapter 3: The Cat's Meow: Learning the Language of Cats 17
The Language of Attention 26
The Language of Compassion 26
The Language of a Name 27
The Language of a Cat Lover 29

Chapter 4: Nine Lives 33
Keep Your Cat's Nine Lives Safe: See a Vet 38
The Kitten 40
    Internal Parasites and Fleas 43
    Upper Respiratory Infections 44
    The Importance of Vaccinations 45
    Distemper 48
    Feline Leukemia Virus (F.E.L.V.) 48
    Feline Immunodeficiency Virus (F.I.V.) 49
    Feline Infectious Peritonitis (F.I.P.) 49
    Foreign bodies, Poisonings, and Accidents 51
    Mites and Ticks 53
    To Spay or Not to Spay 53
The Adult 56
    The Outdoor Cat 56
    The Indoor Cat 57
The Senior 59
    Feline Gerontologist 59
    Cat Heaven 62
    Departing Words 64

Chapter 5: On the Wrong Side of Every Door is a Cat 67
Is The Role of Companion Enough For the Indoor Cat? 67
Cat Obesity 71

The Indoor Cat: Consequences and Real Risks 73
Fleas Indoors 79
The Domesticated Cat 82
Toothaches 89
Environment Enrichment 93

Chapter 6: Curiosity Killed the Cat: The Outdoor Cat 99
Outdoor Cats: The Real Risks of Allowing Your Cat to Go Outside 99
A Note on Dominance and Territorialism 113

Chapter 7: There's More Than One Way to Skin a Cat 117
Elimination Outside the Box 117
Marking 124

Chapter 8: All Cats Are Grey in the Dark 133
The Senses 133
Hairballs 136
The Unseen Parasite 138
Nutrition 141
Live Food 151
Cruelty and Kindness 155
Aggression in Cats 157

Chapter 9: Grinning Like a Cheshire Cat 163
Camouflage 168
Cat Tails 169
Biggest Tail of All 170
The Tail of Three Modern Cats 172

Chapter 10: When the Cat's Away the Mice will Play 177
Good Salesmen 188
T.G.I.F. 191

Chapter 11: Conclusion: The Guru Cat and the Copycat 197
The Energy of a Cat 206

References and Further Reading 211
Author Biography 213

Foreword Dr. Liz O'Brien DVM, DABVP (Feline Practice)

For the last 28 years as a feline practitioner, I have devoted my life to the care of cats. These include cats of all kinds: all ages, all breeds, those with homes and those without, and those that are loved and cared for and those that are not. Dr. Carol Teed and I share this passion—the care of cats wherever they are and whatever their circumstances. Neither of us came to it right away—this caring for cats only. Just as most homes and people that are now owned by a cat, we were lured by the cats. There is just something about a cat—his charisma, his apparent utter lack of need combined with his absolute need for care that clinched it.

In the past few decades a change has occurred and our idea of the cat has shifted. We no longer think of the cat as a loose idea within the family structure. The pampered modern cat is now the most popular pet in North America and they are growing in importance as a family member. At one time, the owned cat wandered his territory in the neighbourhood on his own terms until he was ready to come home for a meal and a nap. Now he is more likely to stay indoors where he is safe from the threats of the outdoors and the complaints of the neighbours. While this new lifestyle keeps the cat safe from the obvious threats of the outdoor life, he now faces new concerns and stresses. Few cat lovers, myself included, ever have just one cat, and as a result, today's cat often lives with multiple cats in the confines of his home. Instead of being free to socialize with his feline friends and family in a society out of doors by the lane or the back porch on his terms and in his timeframe, he is forced to socialize 24/7 on everyone else's terms. As a result, we are seeing a new kind of feline patient in veterinary clinics with new feline health issues.

What has not changed in this same timeframe is that we are still euthanizing thousands of un-owned, homeless, and feral cats on a weekly basis in shelters across the country. Our cat overpopulation crisis is enormous and it is a result of irresponsible cat ownership and a reflection of our society's lack of value of the cat. In this book, Dr. Teed refers to a program called "Care for Cats." It is the first Canada-wide campaign aimed at increasing the value of un-owned, homeless, and feral cats in communities across Canada. As with most campaigns, it has evolved and has now grown into "Cat Healthy Canada." The goal is to make sure that across this country, cats are spayed and neutered, identified and registered, and that they receive regular preventive healthcare in order to improve the length and, most importantly, quality of life. I encourage all readers to embrace this program and make a difference in your community.

This wonderful book, Learning the Secret Language of Cats, delves into the care of cats on both sides of the door. Dr. Teed refers to cats of all kinds: their struggles, what ails them, their resilience, and of course she talks about their language. She states that "every household could benefit from a cat," and I couldn't agree more. The feline-human bond is a magnificent thing, which is good for the mind, body, and soul. Cats are good teachers if we will only take the time to observe and listen. We watch them in balance and in imbalance within their environment and we learn from that. We soon begin to recognize that we too live more on one side of the door than the other and that we too have imbalance in our modern world. The cat lives simply but he does not forget to enjoy life and he is always his true self. Dr. Teed whimsically refers to cats as spheres of positive energy treading amongst us pressing subtly for small changes in the small sphere. I too have witnessed what change can come from a cat in a household.

What I love best about Learning the Secret Language of Cats is its honesty. Carol writes passionately about the language of cats and also the veterinarians and caregivers who care for them. She writes about their struggles, what ails them, their resilience, and the joys of practice, and also the things we usually don't talk about, like the dark side of the street. In addition she shares her own personal journey as a caring veterinarian and feline practitioner.

This book in your hand is not a classic reference book, it is not a story, it is not a memoir; it is all these things. It crosses science and the metaphysical and the spiritual, yet it is all about the alluring, beautiful, and mysterious cat. I believe it is an important book written at an important time to be enjoyed and shared by all. Dr. Teed's honesty, sincerity, and passion are woven throughout this book like the threads in a beautiful cloth, and one can feel the depth of her compassion and love for this beautiful species. It will open your eyes and you will never see the cat in the same light again.

Prologue: Letting the Cat Out of the Bag

The beauty of writing a book about my experience and observations with cats as a veterinarian is that the preparation is already complete. The beauty of writing specifically about cats is that they will provide all the intrigue and drama. There is no need for artistic license here. The material for my book has been aging and mellowing within me, and it is now up to me, after years of working, living, and enjoying cats, to let my experiences flow to the keypad and onto the screen, ultimately creating a book to share with other cat lovers about the secret language of cats. The readers who are cat lovers already know what I am referring to and the others will learn the meaning as they read this book.

My interest in cats began early, when I discovered they liked me. I was the kid that cats found and although my parents would always say it was the other way around, I believed it wasn't. Later, when I first met my husband, he said I was like a cat and he would meow at me when he saw me. I have to say I was quite disconcerted by that, wondering if he meowed at all the girls and I felt it was really quite tacky. I certainly did not wear my interest in cats on my sleeve, so was also a little curious as to how he could have known. I don't even know if I had fully realized it was a passion that others didn't always share. However, my husband continued to pursue this cat in me and he still claims that he feels he is watching a metamorphosis of a "cat woman." In fact, to this day, he still says meow to me at night instead of saying good night.

Working with felines has given me the rare opportunity to observe them from the veterinary perspective through their entire life cycle in a more intimate concentrated way than would be possible if I were strictly a cat owner. And through their accelerated life cycle, at least five times faster than our own, I saw life unfolding in fast forward and realized my own mortality. This was such a gift. And this is also how I began to learn their language. After working with so many cats for so many years, I began to see the cat in a certain way and I want to share that with others.

I believe the cat is fluent in a second secret language and that is the language of disease. It is a universal sign language shared across species.

It is spoken in signs and symptoms—a body language that the cat is quite literally dying, in many cases, to translate for us. Our bodies have the same intelligence and wisdom. Our bodies respond to the environment and lifestyle in the same way that our cats do but we don't often listen or understand it. But when the language of disease is written on our cats it makes sense to read it and accept it as a caution for theirs and our own sakes.

I would describe myself as a "recovering veterinarian" in the midst of going through the steps to recovery. Writing has been especially therapeutic. I think many of my colleagues will immediately understand what I mean by what I have written, as will other cat lovers. Most of us share that intense love of animals, of all creatures; some of us have been called to heal them. It is an intense occupation and many of us succumb to the effects of second-hand disease at some time or other. In other words, working with sick animals every day can bring some wear and tear upon oneself, sometimes on a physical and psychological level as well. Unless you are made of steel, how could you not be affected by some of the pain you see? I am frequently asked why I would give up my practice. It's not something I can answer in a word or two. I know it may seem impossible to many that I would want to. It was the perfect career. For many it seems a waste of so much accumulated experience and knowledge. However, as I have discovered, it is not an easy profession to give up; hands in warm coats, helping those who have no words, and there is always that yearning to take it up again. I have to admit that lately I practice periodically, just to take the edge away, but I am not as involved nor ever will be again; hence, my need to write this book to share with others my experiences and knowledge. But, there is much more to be said about the cat beyond what ails him and how to keep him healthy.

After many years in practice, I did experience a metamorphosis of sorts that probably comes for most people at my age—a kind of meaning-of-life crisis. The cat leapt in and became my teacher. My patients helped me find the answers. Even though I don't know if I am qualified to give anyone advice on how to live, perhaps the cat is. During my years of practice I saw a lot of brokenness in many of my clients, many of my friends, and my extended family, and over time I began to feel it was a universal brokenness that we all share, similar to some of our feline friends. We are all searching for connectedness and also our purpose in this crazy world. And so much of what makes us so confused and unbalanced I feel is environmental, like our cats with their indoor issues. There is a connection, a strong universal or spiritual connection, that binds us all, but we often do not feel it, consumed as we are with hurried day-to-day concerns and worries. As you have chosen to read this book, I would like to think that we share a bond: this love of cats. This book is also in a small part my apology and my explanation for walking away from being a veterinarian and my pledge that I made to heal. It also tells the story of valuable life-changing lessons I have learned through my clients and the many cats I worked and lived with.

Basically, I would like to let the cat out of the bag.

This phrase, "let the cat out of the bag," means to divulge a secret. This phrase was coined in medieval markets where dishonest pig traders would give their customers the piglet in a bag. Once home, the customer would discover a cat instead of a pig in the bag. So, in essence, "letting the cat out of the bag" revealed the secret of the con trick. Today we still use this term. But who is the con? The cat or the human? And what is the secret?

Chapter 1:
The Purrfect Career for Me

It is perhaps bad manners to begin a book about cats without talking about cats. A cat demands that he be centre stage and certainly a book about cats should make him so. Because this book is written through the eyes of a veterinarian, I must set the stage for the cat, the main character, by first briefly showing you the genesis of at least one of the veterinarians that tend them. This is a piece of my journey into feline medicine and places beyond led by the mysterious cat.

Becoming a Veterinarian

Let me start by saying my virology professor in vet school announced during one of our final lectures that he doubted any of us would find true happiness in our chosen field, that we would feel apart, and would have difficulty finding meaning in our lives. I felt angry about his comment because I was accomplishing a life- long dream. Even though I liked this professor very much, I did not let him stop me from pursuing my passion. I did not change gears; I maintained focus even though it was not the first time I'd heard this.

Between graduating with my BSc and entering veterinary school, I worked in an Agriculture Canada pathology lab. I spent my days looking through a microscope at all manner of bacteria: identifying them, growing cultures from tissue and food samples, and counting the resulting bacteria. I see the world a little differently because of that experience. I felt a sense of being outnumbered by a tinier population. Once I began my studies, I realized that our bodies and our cats' bodies are in fact composed in part by friendly microbial populations, our normal flora, which assists us in thwarting unfriendly populations. My time spent with a microscope was the precursor to my more recent growing fascination for smaller and unseen things that make up our own bodies and every other thing in the entire universe.

A few of the veterinarians working at the pathology lab told me the field had been a disappointing career choice for them on many levels; they advised that I might consider research, or teaching, after obtaining my Doctor of Veterinary Medicine (DVM). I smiled and nodded appreciatively but thought the idea was absurd. However, there was a virologist there (yes another virologist, perhaps virologists who study the infinitely small are a little more sensitive to the world at large) and her words caused me to pause. She had given up practice after just three years because she could not cope with the irresponsible pet owners. She came to her decision due to frequent owner requests for euthanasia, which she felt forced to perform on their pets due to pain, suffering, or neglect. She headed back to school to become a veterinary virologist. At the time, I just couldn't understand her reasons for giving up her practice and decided she must have had a unique experience in some closeted cruel area of humanity and that it wouldn't be like that for me.

Certainly veterinary medicine is a challenging career. It is important to fully evaluate your strengths and weaknesses, and also your expectations before deciding upon it. It is not good enough to be the one cats find. It is not good enough to just enjoy animals and want to help them. As many of you already have experienced, you can do that meaningfully in any number of ways outside of veterinary medicine while pursuing other lines of work. I am hoping the stories and experiences I have included in this book will help those considering this career to decide if it truly is for them. The perfect vet has a soft heart and a thick tough skin, and if you have fast reflexes, then consider feline medicine. The rest is nothing more than just hard work and a lifetime of learning.

I don't want to say that there is a particular brand of person that enters veterinary medicine, because my experience was that my classmates came from all walks of life, with different backgrounds and roads that merged to that particular place of learning. Some had lived on farms and worked with animals their whole lives, while others had parents who were vets and knew exactly what they were getting into. The rest of us learned as we went along what it would mean to become a vet. But, I do want to say—because I think it is more or less true—that after some time many of us began to look alike. Of course, this is my own perception, and perhaps some veterinarians would disagree or even be offended by my ideas.

There is an intensity in many veterinarians that does not always match our outwardly easy-going friendly appearance. We don't always fit well into our clothes, needing a quarter- or half-size more or less than what comes off the rack. Even before some of us arrive at work in the morning we are wearing a thin coat of cat hair over our clothing and dog hair weaved through our fabrics. This can give us a disheveled appearance that we constantly fight against because it is at odds with our inner quest for perfection in our work and appearance.

We might feel more at ease in a body suit or fur coat (perhaps tabby) that would help us blend into the surroundings and offer camouflage, so that we may observe unobserved. In essence, we are observers and this skill helps us make our diagnoses. We also enjoy our clients, but this sometimes surprises many people. Some of us don't enjoy small talk and so we crave a deeper connection. Of course, some of us are outgoing and eager to chat, but we are often happy after some time to retreat to our treatment or surgery rooms to be among our patients. Those meditative silent tasks of working with our patients appeal to our natures. The veterinarians that I have the privilege to know have the truest, most gentle hearts of all hearts. Overall, if you asked them what character trait they prize most in a human I think they would say kindness and compassion, which are qualities we share consistently with our feline clients.

Astrology and Vets

I am sure you won't be surprised to learn that after years working with cats I might develop an interest in other mysterious or mystical avenues. For instance, over time I noticed that many of my veterinary friends and workmates that came and went shared the same birthday season—the same numbers. End-of-year birthdays seemed to be popular among us. I was not in the habit of asking clients when their birthdays were, but I wish I had. I would not be surprised if many of my cat-loving clients also shared birthdays late in the year. In addition, I also noticed that some of us share some very basic characteristics beyond our life's work.

According to Gary Goldschneider's and Joost Elffer's book *The Secret Language of Birthdays (1994)*, I was born on the day of the great enigma. An enigma is a person or thing that is mysterious or difficult to understand. It also says that I "have an abiding love for small children and animals; and within these relationships I use a full range of expression through strong intuitive gifts and non-verbal abilities." It says other things that I'm not so willing to share (apparently I use silence as a weapon), but it seems I did not let the astrologers down, even if I did not know or care about it at the time. As a woman of science, these ideas are new to me and I find it fascinating how close to perfect the astrologers are able to list my strengths and weaknesses, especially in linking me to my life's work and love of animals and cats. Looking at it from this far along the path where much of my life has already been played out, I find I can no longer discount it. I wonder if our death day signifies as much.

No matter Goldschneider's and Elffer's claims, I am not so sure I qualify as an enigma, but a cat certainly does. In fact, in my experience, cats are very mysterious and difficult to fully understand.

Numerology and Cats

For fun one day, my Kundalini yoga teacher helped me determine my astrological numbers, which tell me about my soul, karma, gift, destiny, and path, based on my birthdate and year. I have not disappointed the numerologists either. Some years ago when I had less sense, I would have called this nonsense. Now, I'm not so sure. I have peeked through a different microscope—a reverse microscope—a telescope of sorts, and I realize that just as there is so much smallness we cannot see, like bacteria, viruses, and atoms, there is also so much greatness governed by numbers, which we cannot see within our universe.

So, perhaps my colleagues have come from different backgrounds and places but we very likely share numbers, because numbers can be arrived at by many different ways. I have had a fondness for numbers since a very young age so I am not surprised that they mean more than what they appear. For example, the number nine has always fascinated me. The number is often attached to cats: nine lives and the cat of nine tails. The nine times-tables are especially intriguing. The product is always a number that if you add the digits up, it will equal nine. 9, 18, 27, 36, 54, 63, 72, 81, 90. Even 9 x 33=297 and those digits add up to 18, which adds up to nine if you reduce it to a single-digit number, as the numerologists like to do.

This same fascination with numbers overtook me in my clinic when I looked upon my patients and counted their heartbeats and breaths with my stethoscope. The exquisite variety of coat and body types and colours also made me wonder if God may be a mathematician first and foremost. He seems to love permutations and combinations as evidenced by our DNA, our genes; so playful, imaginative, and orderly. I think it is interesting how astrology, and more so how the astrology of numbers, relates to cats and their connections with humans. Some people also believe that the name of a cat (as well as human names) can affect its personality. Basically, the letters, as well as consonants and vowels represent certain numbers that are symbolic of personality traits.

Take the word cat itself. A cat has a six Expression, revealing its domestic nature, but a one Heart's Desire, which shows its independence and often stubborn character. It has a five Personality, which reinforces its independence and need for freedom. You cannot control a cat. Interestingly, the cat has sixes in its core numbers—the desire for love, attention, and domestic harmony. Therefore, while a cat with a six Expression is perfectly suited for the domestic environment, its one Heart's Desire gives it a need for independence. A cat, by nature, is not obedient. I found this bit of numerology wisdom on www.decoz.com.

Numbers and Cats

- 9: *The gestational period of a cat, nine weeks as compared to nine months for humans.

- 15: The human equivalent in years that a kitten will reach at the end of his first year.

- 110-140: The average number of feline heartbeats per minute.

- 244: The average number of bones in a cat's body (about 40 more than humans).

- 32: The number of muscles in each ear of a cat.

- 34: The age of the oldest cat on record.

- 19: Record for the largest number of kittens in a litter.

- 80: The estimated number of known cat breeds.

(Information retrieve from http://cats.about.com/od/factsandtrivia/a/numbersracket.-QHM.htm)
*Items originally added by C. Teed (information about the number nine)

I didn't intend to practice feline medicine only. The cats lured me to it. I can almost see them glancing behind to make sure I was still following as they vanished around a corner leading to my career. Part of my decision to focus on Feline Veterinary Medicine is that I learned that I could manage a cat that others couldn't. I don't know why, but this was a special talent of mine. I would finish an exam and look up to find the owner looking at me in amazement as though I had charmed the cat. Other times when the owner made no admission of previous fractious episodes, I would get a comment from a workmate about how I had managed a difficult cat. I also learned that cats often didn't get the attention that dogs did in mainstream clinics, which were usually very strongly canine based—at

Chapter 1

least during that time, which was some years ago now. I grew increasingly fond of treating cats; they impressed me as challenging and intimidating, and I couldn't get enough of them. I turned my attention exclusively to felines and never regretted it.

Cosmo

Louis

Chapter 2:
Cats are Like Potato Chips - You Can't Have Just One

When I left my practice, one of the gifts that my staff gave me was a little plaque that said, "Cats are like potato chips, you can never have just one," and it is so true.

If one cat in the house is a joy, then two must be sublime, and three or four must be more than that again. I can understand the appeal. We have never had more than three at a time in our house. When one goes, another seems to magically appear within days or weeks to fill the quota. We've had so many come and go over the years, and many of them meet untimely outdoor deaths or M.I.A.s. All of these cats were individuals. All had their funny quirky traits, all different coats, body types, and levels of activity and interests, and I have loved them all. I can spend endless amounts of time staring at a cat like a burning candle in meditation; their coat patterns and colours, the shape and colour of their eyes and nose, their perfectly applied eyeliner, their flicking tail, and chatter when they see a bird. They are beauty itself.

If we provide all their essential needs—both their physical and emotional needs—then there is no reason why we shouldn't have a dozen cats with all different colours or breeds like a moving, breathing Royal Doulton collection. Watching our cats socialize with each other or how they take over a room sprawled in their spots never gets tiresome. At the clinic where we had anywhere from two to five clinic cats plus kittens now and again (when I left there were three plus one of which I was unclear of the status, but likely becoming a clinic cat; oddly every one of them was grey), it was even more entertaining to watch them strut about and meet their schedules.

The selective breeding of cats to produce specific pedigrees is a fairly new practice considering that cats have been domesticated since ancient Egyptian times. Most cat breeds are not more than one hundred years old and there are new breeds arriving on the scene all the time. Our domestic cat is related to larger wild cats such as the tiger, puma, lions, and other smaller wildcats, and they still maintain a certain wildness. There are many small breeds of wildcat throughout the world but most are so timid we know little about them. Our cats probably descended from the more (relatively speaking) friendly and calm African wildcat Felis lybica. As domestication advanced and dispersal occurred it likely bred itself back into the wildcat strain or with other wildcats now and again, and created distinct populations of ancient breeds of cats, which adapted to their specific geographical locations. For example, Oriental breeds originating in hot climates tend to have lithe, slender bodies and long thin tails. Their coats are thinner and paler with less of an undercoat, or double coat, which is found in more northern stockier compact breeds adapted to cooler climates. These first breeds of cats developed in geographic isolation. And it is these specific body, coat, and personality traits found in these early distinct breeds that early breeders sought to cross breed into their own lines to form the breeds known today. Whatever breed you choose to add to your family, you could not possibly find more variety of character, coat, and body type than what is offered from our garden variety of domestic short and longhaired cats found on every corner. Their genetic makeup is no doubt much more complex even than their human counterparts on any corner—a mixed bag.

I believe that in most cases, the limiting factor to providing our cats' needs is space (second would probably be our reluctance to keeping good litter box hygiene). We will discuss this further in other areas but I wanted to mention it here because I know the addictive nature of cats, like potato chips, and know many of you will have that one-more-cat feeling from time to time. Overall, cats are certainly addictive.

We have to consider the reasonableness of our space before adding more personalities into the household. We must consider it for the good of all before we tip that balance for good. Space matters less of course when a cat is permitted time outdoors where space is limitless. Inside a household, there needs to be space for litter boxes; space for sleeping; space for hiding, for quiet time, and for avoiding or escaping conflict; space for sunning; space for playing and exercise; space for viewing the great outdoors; space for eating and drinking; multiplied by however many cats you have. Space can be horizontal or vertical. One great way to increase space is actually to provide vertical spaces like ledges, climbers, perches, cozy chairs, and more chairs. What else do our cat friends need from us? A sturdy carrier, food and water bowls cleansed daily, litter boxes cleaned or scooped daily, a variety of toys, beds, blankets, combs, brushes, nail clippers and other grooming supplies, hairball remedies, quick-release collars if they go out, microchip identification, sturdy scratching posts, cat doors, things to climb on, and if he is indoors-only then an outdoor enclosure of some description is recommended.

A single modern indoor cat can be perfectly content. Some cats actually do better as single cats. I think for most households, two is a very good number. For me, three is perfect—or almost perfect. Watson will certainly tell me if the boxes aren't clean enough, or if the dogs are rowdy and in his face; he tells me these things by peeing on a bed. So I need to manage his environment carefully. Louis also has his long-term defecation out-of-box problem but he already had that when he came to me from living with five other cats and two large dogs, and the space was not managed and was far too limiting. I fear that the other two cats marginalize Merry sometimes, but she has become more and more social in the past year, especially now that two of our children are off to university. Merry's favorite child is one of the absentees, and she is over the moon when she comes home, but I think she likes the extra space with fewer individuals. I think also she is finally becoming more comfortable with Louis after the two years he has been with us. It is a balancing act. Space and numbers, human and fur coated, is key to balance. Personalities also matter. Some personalities take up an enormous amount of space, but then again, the same can be said of some humans.

When considering enlarging your cat family, individual temperaments are the most important consideration, but unfortunately it takes time for a cat to settle into himself in a new home, and we might not know at first glance what sort of cat he will be. Adding a juvenile to an established house of cats makes the most sense, since adult cats will tolerate a youngster more readily than another adult who already has ideas about his own dominance. However, even when bringing a kitten or young adult cat home, there will be a few uncomfortable days, and the time the cats spend together should be brief and supervised. Once the adult cat understands that the kitten is not a mouse or a rat, but likely just a silly disrespectful kitten, things will go well. Our clinic cat Beau loved kittens.

He took every one of them under his wing and mothered them. It was so fun to watch him follow them around, groom and play with them, and sleep with them curled up under his protective arm. It certainly can be a thing of beauty. But there was always a day or two of distain and watchfulness before he gathered them in.

I don't know if sex matters so much; personality matters more, but I think two male cats (neutered of course) get along best, or a male and female pair (neutered and spayed of course). Two females can get along wonderfully but I see more problems with this combination. Siblings are especially nice; they grow and play together, keep each other company, and are double the trouble. A pair of kittens tends to be more cat-centric; in other words, they may become their own clique and less interested in the humans in the house, but that generally changes as they get older. Social maturity can bring changes to siblings, which sometimes makes them less amiable toward each other later on.

Sometimes we can't choose the age or sex of the new cat in the house. We see them and they have to come home with us for whatever reason—perhaps true love. Sometimes they choose us. And then we just deal with it by managing the environment best we can. Sometimes we are combining our human families and there is always stress about whether the pets will get along.

Blending families can be done over a period of time. Remember they are already familiar with each other's scents, so they've had an introduction of sorts from your clothing and person. Swapping scents with a towel rubbing is helpful or even just bring extra articles of clothing from each home to help acclimate them further to the scents. Swapping food bowls and even gently used litter boxes can also help before the big introduction. Allow each the opportunity to explore new surroundings by themselves to identify each other's scents in situ and to locate resources like feeding areas, elimination areas, sleeping and hiding areas, and other points of interest. This can be done with the pets separated by closed doors first, and then by having them together for brief and supervised periods of time while possibly on leashes, but this is not usually necessary. Using pheromone sprays or diffusers for a few weeks can ease tension. Eating together is a wonderful way of making a start to their integration because they will be focused on their food. Let them get hungry, keep the bowls on opposite sides of room, and bring them closer each day. Feed them something extra tasty. This is a good time to give treats.

Remember not to inadvertently praise poor reactions, which will just reinforce and perpetuate more poor reactions. Provide extra support and love during these early weeks, but be careful not to cuddle and pet them in response to reacting badly to each other. Instead, separate them and begin again to introduce them in a day or so. There is no need to hurry introductions; let them acclimate slowly to each other on their own time. It has taken Merry two years to be comfortable around Louis. And oddly it has taken the same amount of time for Watson and Louis, who are both quite dominant cats, to move out of mere tolerance for each other to the formation of a brotherhood where they sleep swirled together and groom each other until they are wet. This has just happened within the last two weeks. We've sent our daughters pictures to prove it because neither of them believed it. So, there is always hope, but you need to manage the space.

A few practical considerations before you bring your new cat (for now, let's call him Chip) home. It costs twice as much to have two cats as one. It costs four times as much to have four. More food, more litter, more litter boxes, more toys, more bowls, more boarding fees if you travel a lot and tend to board your cats, and more veterinary expenses including routine spaying and neutering and vaccines and more veterinary expenses related to illness. If the idea of more insurance is unpalatable to you, then consider starting some kind of fund for your cats' care—call it a kitty fund if you like. Make monthly contributions for each cat and decide if it will fund routine care, or just illness and emergencies. I had several clients with large numbers of cats. I can think of one in particular who could not say no to a cat and they seemed to find their way to her through some inter-cat news network. They would basically just show up at her house looking for dinner with a, "Sorry I'm late," kind of look on their face and would decide to stay forever. I think this dear soul spent more money on cat food than her own food. She also had several cats with serious health issues so she was a frequent visitor at the clinic with one or another of her cats. It must have been nearly a full-time job looking after her many cats with their regular daily needs and special needs. Sadly, many of these cats were close in age and although in her wonderful care they lived very long and full lives, they began to die one after the other and this was a very difficult few years for her.

Hopefully, Chip will not suffer a sick day in his long life, but I worked day after day for 23 years in veterinary clinics and can tell you that the chances of him becoming ill at some point and needing a veterinarian's help are real. You do not want a bad month of financial woes to make the decision for you, whether you treat him or not. Illness is rarely convenient.

Before bringing Chip home, make certain that your resident cats are up to date on at least their upper respiratory vaccines. Make sure Chip is screened for internal parasites, fleas, and ear mites. Make sure Chip has had a full physical exam, is vaccinated and dewormed, and is tested for Feline Leukemia Virus and Feline Immunodeficiency Virus at a minimum. Your veterinarian may also recommend a routine blood panel and urinalysis to look for any underlying disease. Older cats may have some dental issues that need to be addressed right away or other age-related issues such as arthritis or kidney changes.

These steps are to protect your brood already at home and also you. You will quickly become attached to your new feline and you need to know if there are issues up front that need to be managed. Many of you will laugh when you read this because you know as well as I do that it is already too late. If he has stepped foot into your house, you already love him. It happened sometime between putting him into his carrier for the first time and releasing him into your home. Or simply when you opened your door to the stray that's been hanging out for a few days or a few months. It happened when you realized that he chose you.

If after examination Chip is found to have dental disease and needs painful decaying teeth extracted, or if his blood work and urinalysis show he has kidney compromise, then at least you will know ahead of time what you are taking on. You will probably still welcome Chip into your home, but you now know what that fully means and can feel even better about giving him a home.

Another concern when bringing a new cat home is the possible consequences to the health of the humans in the household. We actually don't share too many things with felines, but there are some risks that you need to be aware of. For example, we cannot catch Feline Leukemia

Virus or Feline Immunodeficiency Virus. Fleas will bite us but will not stay on us, though a home infestation is most unpleasant. Yersinia sp. can be carried by fleas and can infect humans, but this is rare. Lice are species specific and will not bother with us. There are several parasites that we can contract from our felines, but good hygiene goes a long way to prevent that from occurring. It is more likely in the very young and the very old or if our immune system is fragile. Transmission of these parasites is generally through the oral route after contact with infected fecal matter. Giardia, Toxoplasmosis, some roundworm, helminth, and tapeworm species can be passed to humans. Many cats carrying these parasites will not be ill and may also have normal-looking stool. If a cat has certain bacterial diarrheas, such as Salmonella, Campylobacter sp., or E. coli, then they could be passed on to humans. Prevention involves good hygiene practices after handling your cat, good deworming practices, fecal float tests to detect giardia and other single-cell parasites, and careful daily cleaning of feces from the litter. Follow your veterinarian's advice on deworming protocols. Diarrhea fecal tests including floats and cultures for example, often need to be done to find the cause so the cat can be treated. Leptospirosis is another bacterium that can be passed to humans but is rare.

Ringworm is a very itchy fungal skin infection that we can obtain from an infected cat. It can have a lengthy pre-patent period, or period between exposure and a lesion developing, so sometimes we don't know where we have contracted it. These lesions can be unsightly, especially if they form on the face. They are intensely itchy and typically quite round as their name suggests. They require a visit to your physician and special creams to treat the fungus.

In general, bite wounds can cause serious infections in humans that often require antibiotic treatment and medical attention. Scratches can also cause serious infections. Cat Scratch Fever is a bacterial infection primarily caused by Bartonella henselae, though other organisms have also been implicated. Humans, usually children, may become infected by a scratch or bite from an infected cat. Kittens are most likely to be infected but up to 40 percent of all cats have carried this bacterium at one time or other, usually with no health concerns to themselves. Infection of the bite or scratch area may be followed by lymph node enlargement and flu symptoms up to two weeks afterward. A physician should always see bite wounds and scratches that appear swollen or tender. Again, immune-compromised individuals may be particularly at risk for developing complications related to Cat Scratch Fever. Preventing transmission of this bacterium can be done by not engaging in rough play that would elicit biting and scratching in kittens, washing after handling cats and kittens, and not allowing the cat or kitten to lick sores that you may have on your skin already. Rabies is another deadly virus that can infect humans and is generally transmitted through a bite wound that breaks the skin. There are other zoonoses (transmittable to humans) that can occur. It is always best to consult your veterinarian on all potential zoonoses.

An overly active cat may also demand a schedule that you are not prepared for. A dear client of mine adopted two adult cats from a client who had died, which left her cats needing to find new homes. The clients fell madly in love with these two cats, but their sleep schedule was turned upside down. They had to start going to bed earlier to accommodate the cats waking them quite rudely every morning at 4:45 a.m. There were no more leisurely mornings drinking tea in bed for this retired couple. This was not exactly a health concern for the couple, but had they been cater-wauling cats in the middle of the night, it might have been. This happens especially in older cats with cognitive dysfunction or hyperthyroidism, or of course in unneutered males and intact females in heat.

Again, this brings me back to the saying, "cats are like potato chips." In essence, we have one, and then we want another and another. However, too many can be unhealthy. We need to consider all the consequences—good and bad—that can come with any new addition, including the meshing of personalities within a limited space, added work associated with litter management and general housekeeping and grooming, added expense in veterinary care and feeding, and health concerns that may exist already and may come with advancing age. If you are concerned that your single cat needs company, don't be too sure. Many cats in loving homes enjoy their single cat status where all efforts are focused on them. I would go so far as to say that many cats would choose this lifestyle. Sometimes, one chip or one cat is often enough.

Chapter 3:
The Cat's Meow: Learning the Language of Cats

Apparently the phrase, "The cat's meow," arose in the 1920s to describe something especially fabulous. To me, cats are fabulous. In turn, I believe that if you learn the language of cats, this will help you understand them better, and appreciate them even more.

Learning the language of cats is a lifetime pursuit. I have come to believe that it is a worthy pursuit. It is of little wonder to me that the ancient Egyptians held cats in such high esteem. I believe that cats have so many lessons to teach us if only we will watch, listen, and learn their language.

Meowing is actually a specific mode of communication between the kitten and mother cat. Meows granted to humans are an extension of this language. Cats rarely meow to each other, for example, or to other animals. Realizing our inferior body language skills, they grant us the meow. It may be in greeting but usually it is requesting something; attention, food, or an open door. The meows of cats can be as distinct as any human voice with special pitch, intonations, expression, and use of syllables. Most cats have several different calls that they use, which mean slightly different things. Cats tend to meow as they approach you, which I always thought added to the drama of a cat.

Our most expressive cat is Louis. When he is lonely and looking for company, he has a meow that sounds like a questioning, "Hello?" When they are hungry, Louis, because he is not afraid of the dogs, is sent to alert me of this situation and that meow is a series of short staccato, "Ma-mrrra"s—the last one almost spit out of his mouth for effect. He's saying it is urgent and he doesn't want to ask twice. Or perhaps he is just worried that Watson will cuff him if he doesn't arrive back with me following behind him. When you are a cat, it is a very treacherous thing to be in a house with two large dogs during the first hour or so in the morning. Their jubilation and celebration over the rising of their humans can easily turn into a cat chase, or worse, turn into being cornered. It is a very great humiliation for a cat to be cornered by a dog (or two) with which he shares a house. The only thing predictable about a dog in that embarrassing state is how unpredictable they can be. I can picture the cats gathering each morning to decide which cat will hazard it today. Louis of course is the obvious choice.

Our cats cannot be fed free choice because Watson, who abuses every liberty also abuses the idea of a full bowl, and so it must be metered out into several meals so he does not get more than his fair share. When I arrive at their bowls, Watson chastises me for taking so long. Watson starts with Louis' bowl, and then moves to his own when I pat his bottom. Then Louis settles in to eat but he's not in any hurry. Merry spins around me and will not eat until she has had a good body massage by my fingers down her sides deep in her coat.

Purring is an interesting vocalization specific to cats. Large-breed cats that roar like lions, tigers, and leopards can also purr but only on their exhale, which is unlike our domesticated cat, which purrs on both inhalation and exhalation. It is theorized that purring occurs when air passes over the vocal folds and muscles of the larynx, which causes alternative dilating and constricting of the glottis during inspiration and expiration. The diaphragm becomes engaged and produces continuous sound, though it is usually louder on the inspiration segment. Kittens begin to purr at two days of age and the sound and vibration becomes an important mode of communication between the kitten and mother cat. Cats generally purr when they are content, but they will also purr to calm themselves when stressed or frightened, and when they feel unwell. Elizabeth Von Muggenthaler, a Bioacoustician, says the cat purr is exactly in the range of frequency of 20-50 Hz, which is known to produce healing, like a built-in pain relief and healing system, which is amazing. I suspect a cat on your lap will advance our own healing. The yogi contends that the Ohm sound does the same thing. I am intrigued by this idea. Eighteenth-century writer and philosopher Novalis wrote that, "every illness is a musical problem—its cure is a musical solution." I wonder if he owned a cat. He was likely being poetic, but there is probably more truth in it than he realized.

There is an old veterinary adage that says, "If you put a cat and a bunch of broken bones in a room, the bones will heal." Tissues and organs resonate at specific frequencies and so it is thought that sound frequencies matching these can speed healing. The frequency of the purr is especially good for bone and ligaments—hence the cat's ability to heal so well from falls. But it is also good for other tissues. In a world that is becoming increasingly wary of medicine, it is interesting to me to see renewed interest in the cat, but as a healer this time. Sound therapy, like laser therapy, is gaining recognition for healing. I am a believer in the power of the purr because I have seen it with my own eyes and so I know there must be merit in these other modalities. The cat, being flesh and blood, has an added talent for healing that an object cannot match. That is the power of positivity that he radiates into any room he enters. Therefore, a cat's very presence is healing.

See below the Phonetics of Feline Vocalization, which I found in Bonnie V. Beaver's *Feline Behaviour: A Guide for Veterinarians* Second Ed. Saunders Publishing, 2003. The original source is data from Moelk M: Am J Psychol 57:184-205, 1944.

Murmur patterns

1. Grunt
2. Purr ´hrn-rhn-´hrn-rhn
    a. Greeting (request) ´mhrn
3. Call ´əmhrn
4. Acknowledgement ´mrhŋ

Vowel patterns

1. Demand ´mhrn-a´:ou
    a. Whisper ´mhrn-ɛ´
    b. Begging demand ´mhrn-a:ou
2. Bewilderment ´maou:?
    a. Worry ´mæ ou:?
3. Complaint ´mhŋ-a:ou
4. Mating cry (mild form) ´mhrn-a:ou
5. Anger wail wa:ou:

Strained intensity patterns

1. Growl   grrr...
2. Snarl   ˈæ:o
3. Hiss   ˈsss...
    a. Spit   fft!
4. Mating cry (intense form)   ˈø-øˊ:ə
5. Scream   æ!
6. Refusal   ˈæzˊæzˊæ

Key
(a) as in father, (æ) as in cat, (ɛ) as in get, (ə) as in momma, (o) as in go, (ø) as in French eux, (u) as in pool, (f) as in fan, (g) as in gone, (h) as in hunt, (m) as in mouse, (n) as in kitten, (ŋ) as in sung, (r) as in rat, (t) as in cat, (s) as in see, (:) as in prolongation, (~) as in nasalization, (ˊ) indicates stress-accent, (ˋ) indicates inhalation, (?) indicates rising inflection, (z) indicates wavering or discontinuity, (!) indicates abrupt, stress-accent ending.

Further to meowing and purring, cats will chatter when they watch birds or other prey. A silent meow (from the newborn kitten for instance) can be heard by other cats, but is too high in frequency for our insensitive human ears. A cat vocalizes upon both the inhale and exhale. Cats who are spoken to often will generally be more talkative and may even attempt to mimic certain sounds. When they observe other cats outside, they may grunt, huff, or even scream or snarl. A hiss or spit may do depending on the situation.

There is a fair bit of meowing in our house—the feline focus is not far from home. It is almost a second tongue. Our cats are very vocal, partly because we talk to them so much, but I'm afraid we humans in the house have adopted the language as well to some extent. Our youngest daughter, whom we call Kitten, protests she is becoming a little too old to respond to, "Meow" or, "Here kitty kitty kitty,"—a game we have played since she was a young toddler. Then again, cats have always surrounded both my personal and business worlds, so it is no surprise that my children would become part of it. The language of cats goes far beyond the vocalization of their meows, trills, chirrups, hisses, and purrs.

Chapter 3

They can say much more with their subtle but effective body language.

For example, when a cat stares into your eyes and gives you a slow partial or full blink, then that is as close to an, "I love you," as you will ever get. It is like a passionate kiss. He may also show his admiration with an extended paw to touch you as he sits close by, and head butts and side swipes with his cheeks marking you as his own with the scent glands found there. Other signs of contentment are kneading or pawing their humans or materials they sit on, usually while purring with their whiskers forward and erect. This language goes back to the language shared with their mothers. The kneading stimulates the mammaries, which of course then brings milk and comfort. Kneading also serves as a scenting mechanism and claims you or whatever they are kneading as their own. They may also lick you as though they are grooming and caring for you.

Rolling over is a playful posture or may be a greeting or show of submission. They may lay on their backs exposing their underbelly in total submission to you. Touching noses is also a compliment and a sign of affection. An open mouth shows interest or playfulness. An open mouth with the lips drawn back and the head slightly raised tasting the air is called flehmen. The cat does this to taste the scents of other cats that may have been in an area recently, and to understand them in a second sense. This is something we observe quite frequently in a feline practice. My cats also did this regularly when they inspect me at the end of a day when I arrive home. They sniff me head to toe and then proceed to taste the scents through flehmen. I am sure they learned to recognize some of my more regular patients.

Signs of discontent are also very readable amongst cats. A cat's ears can indicate just how upset he is, whether they are at half-mast or fully down. The cat pose associated with Halloween, where the cat has an arched back, erect fur, puffed whiskers, drawn-back ears, arched tail over the back, and puffed body turned to the side, is a defensive pose. If distance is not achieved by this stance, then the cat may crouch or even roll over with all four paws ready to defend himself, or he may back away very slowly with his tail curved down in an upside-down U-shape. He will not attack unless there is no other option. An offensive stance by a dominant cat ready to attack is a forward stance with ears and whiskers forward and an intense stare with pupils constricted. Back-and-forth movement of the tail, especially the tip, is a gauge to how serious the situation is. Less serious offensive behaviour may involve a subtle turn of the ear, or a flick of a stiffened tail can alert a subordinate cat to stay clear.

The cat's tail has its own language. It sails high when happy, twitches at the tip when excited by watching birds; for example, when hunting, the body crouches and the tail swishes slowly back and forth and low to the ground. The tail will go to half-mast or fully down when disturbed or unhappy. When a cat is about to attack, it will wave itself slowly from side-to-side, gradually lower until it is perpendicular to the ground. Veterinarians watch a cat's tail in the exam room. It is the best indicator of a cat about to strike, and acts as a warning. Learning the language of the cat is what saved me from any serious wounds during my years of practice—that and my gifted cat-like technician who helped me with all my more challenging patients. My eye got scratched once, which often makes me think how precise a cat can be when he strikes. He could have sliced it right open even with my fast reflexes, but he didn't, and once he'd made his point with a final turn of his back and ejection of his anal glands (also perfectly aimed), I had no trouble continuing with the exam. His nails had grazed both my upper and lower lids so that I had an impressive quadruple scratch down my face for weeks. It was a reminder to not take anything for granted when meeting a cat in an exam room for the first time. It was very well said.

The language of a cat consists of more than their subtle but effective body language. More so, the language of cats is learning about the way they live. For example, they understand as much as they need to. As you watch a cat prowl the great outdoors, roll in the garden, or nap in a sunbeam, you know that he is in awe of all of God's creation. He lives in the moment. He smells the roses. He admires a butterfly. He enjoys life. Every day is a fresh day. To understand this about a cat is to start learning their language—the way a cat thinks and lives.

What is also interesting to me about the language of a cat is that the cat chooses. And he chooses within seconds. He chooses the vet, and he chooses the treatment. A cat assumes the best in everyone, allows the first few seconds for you to show your true colours, and then the cat makes a decision. I observed cat decisions day in and day out, year after year. I never took it for granted. They continued to choose at each

Chapter 3

encounter, with each treatment suggestion. I respect that in a cat: their willingness to engage, but also the precision, fairness, ability to know his own mind so well, and his integrity in sticking to their beliefs.

I love that cats always assume the best and rarely entered my clinic with preconceived notions; shouldn't we all be like that? It is often the case that cats exhibit profound fear aggression; they fear a loss of control in the veterinary setting. They should be forgiven for making hasty judgments and displaying aggressive behaviour. Cats will generally grant a second chance once they've felt your energy for themselves in the space created between their breaths or their growls. They can't be talked into cooperation; they respond to intuitive intelligence, not verbal intelligence. They listen to body language and read the quality of the energy surrounding us when we're near them. This is where it pays for a veterinarian to know the language of cats. However, at times I grew to regret that I had charmed a cat by degrees into accepting treatments that I knew would alleviate pain and suffering, but would also extend his natural life span to include further trauma or treatments.

In my years of practice, I also grew to realize that there are those people that love cats and those who don't. Therefore, most cat lovers learn the language of cats over time, whereas people who don't like cats never learn the language, and consequently, never understand cats, which furthers their dislike. In turn, the cats sense their lack of understanding and typically do not trust these individuals.

I have personally developed a general distrust of individuals who claim to not like cats, and there are many of them. It is embarrassing how vocal they can be on this point, and they clearly don't see that it tells something about them. I am not just talking about bad manners; perhaps I was oversensitive since declaring their distain to me, a feline veterinary practitioner, basically nullified my worth to them as an individual. I began to understand that many people have dispensed with their own intuitive intelligence and may feel uncomfortable around cats and humans who depend heavily on their powers of observation, feel, and intuition.

Still, for me it is difficult to understand why some people do not like cats. I have several friends who have severe allergies to cats. One friend in particular cannot even come to my home without first drugging herself up with allergy and asthma medications. I wonder how she can still claim to like cats. I remember going to a wine-tasting party directly after work some years ago. I did not have time to go home so I brought clothes to work with me to change into before I made my way to the party. Once there, a gentleman several layers ahead of me in the throng began to sneeze violently, which upset his wine and everyone else's. Within minutes he was accosting everyone to his right and left angrily accusing them of smuggling a cat into the winery. I had fresh clothes on but I was probably oozing cat all the same having just come from work. Kisses and cuddling are part of every treatment plan and I knew I must have been the cause, as did everyone else who turned my way and just stared at me. Guilty! Some people do get violently ill from allergies to cats and other things and I do not blame them for feeling angered by it. It is a terrible affliction and would do nothing to promote a love of felines. But for others, what is there not to like? Do some people fear that a cat will tell their secrets or exhibit distain toward them? They are so sensitive to the nature of others—cats do tend to intimidate, don't they? I've come to think that people who don't like cats just haven't really spent time with one yet or that they haven't learned the language of cats. In fact, I believe that there are those who love cats and those that don't love cats yet. Perhaps once a cat begins to show a human how to use their intuitive intelligence and understand the cat's language, then they'll become a cat lover.

Practitioners notice this process (the lack of understanding cat language) happen again and again in veterinary clinics. For example, take the stubborn, crusty, abrasive, older man who has been given a cat for company by his well-meaning daughter. He comes into the clinic embarrassed to be seen with a little kitten, even in front of us who love cats. He insists, with his big man arms, that he doesn't really like cats. It is often apparent that he's a little bit angry too. Over time, as vaccine appointments come and go, the number of photos that arrive with the kitten to show us his antics increases, the number of reassuring caresses and kisses that he gives to his kitten increases, and the crustiness in his personality dissolves with each successive visit. It is a marvelous thing to behold. That same stubborn older man has already learned the value of a cat. He has learned that the purr, the vibration resonance that runs through the kitten and through our very selves, is healing; it heals those little tears in our energy fields, and it heals our auras that make us feel

Chapter 3

unbalanced. The kitten tells us to sing and heal. The cat stretches and runs—it feels good, and we do the same without realizing that we are following his example. He washes his face and straightens his hair and we do the same. And we feel better for doing it. Or, at least we may tend to follow their example. I learned lately in a yoga class that the child's pose with the forehead firmly planted on the ground is grounding and will relieve stress and it occurred to me that I have watched cats do this. I should have known that all those heads pressed into my face and my chest and against my legs were telling me something that I needed to learn.

As the years of my practice sped by, I learned not to take offence to the, "I don't like cats," statement, even though I did not discover a way to respond other than with kindness. I smile inwardly and think to myself, Not yet. It is always best to assume the best in others, so says a cat, and I agree. We are all equal in God's eyes; some of us just see it a little more clearly. For those that lack clarity, it seems a shame they have not found a way to wonder at the marvels of a cat; their intelligence, grace, beauty, gymnastic abilities, precision, lack of self-consciousness, and their otherworldliness. They live well, and they die well, with equal grace and beauty. And everything between those first breaths and their last—a life lived on a different time dimension than our own—could be a guide to us on how to live our own lives.

Unlike humans, cats have not lost their instinctual responses, and it makes me wonder if they may not be just a little bit more in sync with all of life because of it. Perhaps this is because cats don't have the power of words to lure them away from their intuitive intelligence. Humans have learned to trust only in the spoken and written word as their truth and have given up other ways of gathering information and learning the languages of creatures that are not human.

I have found my stethoscope to be a wonderful tool for tuning in to the sound of the beat of life common in us all. It would be difficult not to honour each of God's creatures after that. Of course we know there is a heart in there in each of us, unseen, unheard, but hearing it has impact. We are each equipped with our own timer—our own drum. I have discovered that meditation has that same tuning in effect as my stethoscope, which is a grounding effect. I learned that from a cat.

The Language of Attention

Have you ever noticed how cats tend to appear when you settle down to read? My cats are especially fond of newspapers; as they climb onto my lap and foil my attempt to read, they are fascinated by the way the paper crinkles under their paws and how the paper folds down. I pick a paper up very quietly so as to not invite attention, but inevitably I hear the thunder of paws coming to investigate what is taking my attention away. My cats don't care about my attention until it is elsewhere. It is usually Watson (but both Merry and Louis take their turn at this and even Beau at the clinic) who cannot tolerate a paper between us. It is not just newspapers. It is basically anything that removes my focus from the cat, incase they need it. I nearly always type over the body of a cat as I write at home. At the clinic, Beau sprawls across the keyboard of the computer; he does this so often that at times I would have to close my door because I don't have the time to lose another document to Beau's typing. Beau also hung up on several clients with a step of a paw. He was a very involved clinic cat.

These descriptions above are not so much of a cat underfoot, but more about our intertwined worlds as cats and people. The cat that suckles our clothes and soaks them through; the cat that marks our things; the cat that carries our underwear around the house; the cat that shreds his nails by scratching the door when we leave; the cat that will only eat if we give him food out of our hand at the dinner table; the cat that must come into the bathroom with us, but looks the other way out of courtesy, or doesn't look the other way out of attachment; the cat that nips us when we get up to leave the couch—all of these cats have developed stronger-than-normal attachments to their caregiver. This then stresses the importance of knowing what your cat is telling you by knowing your cat's language.

The Language of Compassion

During times when a cat settles into our lap and we reluctantly accept them being there, it is because we may not be sure that we are worthy of the attention. Perhaps we may not want to break the spell of our own misery. And then the cat begins to purr for us because he knows it feels good and he likes to do it and he knows we cannot do it for ourselves. That purr has healing power like a prayer or a chant. That is the compassion of a cat.

Example of How a Cat Can Keep a Sensitive Child Company

A cat will also be a wonderful friend to a sensitive child who has not quite figured out that this sensitivity is actually a gift and strength. A cat will sense the sensitivity of a child. The child may learn to cope better with this sensitivity during therapy sessions with a cat, which are unstructured, but no less helpful and meaningful. Or maybe the sessions are structured; how can we know for sure? The confidence of a child may bloom while taking care of and interacting with a cat. Thereafter, the idea of making friendships with other children may be less daunting to the child.

The Five Steps of Naming Your Cat

1. Get to know the cat before you actually choose a name.
2. Look through some baby naming books or websites for inspiration.
3. Alternatively, try translating words that describe your cat into other languages using dictionaries or online translators.
4. When you find a proper name, call your cat with that name.
5. Call the cat by its name when you feed them, pat them, hug them, and call them.

(Source for the above information: www.wikihow.com/Choose-a-Name-for-Your-Cat)

The Language of a Name

Names are part of our language and they inevitably become part of your cat's language too. A cat knows its name and typically comes when called. In essence, the name you choose for your cat is a language you share. What is in a name? Perhaps not much, but I tend to think there is a lot within a name. Whatever name you choose, it is very beneficial to teach your cat to come to his name by using treats or whatever means works. This can help bring him home if he happens to be out, or can even distract him or call him away from something he should not be doing.

In the beginning, a name tells more about the person who decides upon that name than the cat that is named. Certainly, the longer we wear a name the better it fits. This phenomenon is easily observed in veterinary clinics. For instance, a cat by the name of Killer or Demon (who was once a kitten by the name of Killer or Demon) would be extremely aggressive. When I saw that name, I did not need to see the cat to know that I would need all my wits about me, and also need an extra misting of pheromone spray as my perfume as well as my fast reflexes to manage the appointment. And I knew that any incident that drew blood, like a slash of my wrist, would be hilarious and bring elation to the owners. It would be like an extreme fighting match in the cage of my exam room with the audience goading the cat on to go for my throat.

Now maybe this kitten was named Killer because he was a feisty kitten, but the name becomes a self-realized prophecy. I think that if Killer had been called Rose, then I would not have needed gloves. That is not to say that I have never seen a Rose that challenged my reflexes; only that I never met a Killer that did not.

I was at least as concerned, sometimes more, when I saw my next patient was named Dipstick, Phlegm, or Roadkill, or names that I cannot say aloud. With names like these, I knew that there was not much respect there and that the cat would very likely have fear aggression developed from challenges at home. I think it may be true that saying a name a million times carries some charge and may reflect or even determine not just how we feel about the pet but also how we treat it.

For female cats, I prefer the names of flowers like Petunia, Lily, Marigold, or Dandelion. Or names of vegetables that are pretty or resemble flowers like Spinach, Celery, or Broccoli. Or names that sound almost like flowers—Daphne, for instance. There are so many beautiful names. Why not pick a name that makes us happy a dozen times a day?

For males, I was always attracted to good solid no-nonsense boy names like George or Dan. What I did not like so much in practice were names that were unisex, like Fluffy, Peanut, or Tiger. Climby, Scratchy, or Stinky were other names that kittens were given by the little ones in the family, but generally a parent could manoeuvre a name change before the next visit. Unisex names get a vet in trouble. As a vet, it is difficult to make it through an appointment not remembering the sex of a pet. You can keep saying his name over and over in discussion but eventually you mess up and say either her or him or he or she and the reaction is far worse than saying the wrong name. It is a very great insult to call a cat by the wrong sex. It is almost unforgiveable. It is something you have to work very hard to correct to reestablish your good name. My girls used to highlight the cat's name in either pink or blue to help prevent loss of clients through mistaken sex. It helped a great deal.

I had a bad habit of calling cats, "Kitten," which also got me in trouble. I used it as a term of endearment of course, but clients were always quick to correct me that they were full-grown cats, already 5 or 10 or 15 years old, which of course they felt should have been obvious to me.

I also had a bad habit of forgetting client's names or calling a client by his cat's name. Had they had names of flowers or pretty flower-like vegetables, then I would have done better. I would remember a Mr. Rose or Ms. Cabbage. I had one particular client whom I was so fond of, but I kept referring to him as Mr. Cuddles. He became Mr. Cuddles to me, and I don't think he minded so much.

The Language of a Cat Lover

As a veterinarian who specializes in felines, I had the rare opportunity to observe and engage with many cat people. Ailurophile is the term used for us cat lovers. Working several years in a conventional clinic and also in an emergency clinic before opening my feline practice, I can say without hesitation that cat lovers are a group unto your own. And I think you are amazing. But not because you love cats—it is because they love you.

Cat lovers can be found anywhere and everywhere. For example, my husband once worked in the motorcycle tire industry and I found myself at a function with him once where leather and bandanas seemed to be the most comfortable and natural attire. Our table consisted of several large burly tattooed men and their leather-wearing wives. It was the men at that table who began to tell me story after story about their pampered and well-loved felines. One was a photo-carrying member of the cat lover's society. He produced a photo of his very handsome brown tabby cat named Harley, and explained that Harley purred a lot; perhaps he felt as though he needed to explain this with the name he had chosen. Because I could only see my own eyes in his large mirror sunglasses above his handlebar moustache and beard, I was not sure if I was allowed to laugh. The cat was wearing a miniature bike helmet. I wanted to ask if it was a Halloween costume, but I did not. He said very sweetly and softly that Harley had passed away recently from Feline Leukemia Virus. I did not expect such softness from someone so outwardly tough. I smiled for weeks at the thought of that evening.

Even now, I can picture a cat person standing across from me on the other side of the exam room table with a beloved cat between us. The cat has already made his decision about me, but the cat person remains reticent. It may take a few more minutes or even a few more visits to the clinic before the cat lover decides for sure. It is observable in their eyes and body language and in their hesitance to trust, and I know that I'm the same as they are—I understand them. Perhaps many people don't understand us cat lovers and we suffer for that at times; our reluctance, our tentative qualities, and our intuitive intelligence may be intimidating. We are extremely articulate but use an economy of words; we are precise. We are as close to a cat as can be, and that is why cats choose us. We ask the difficult questions, we know half the answers before we ask. We look for better than average. Cat lovers challenge me to be my best. And once they've made their decision, they become fast friends with me. I see something in their eyes and know that they, like me, feel too much. It is these qualities in cat lovers—that extra sense, that sensitivity—that attracts the cat. We cat lovers communicate with intuitive intelligence, like cats do.

The language of the cat, the cat's meow, is not a secret. It is wide open for us all to learn. Their language contains all the elements of how to live a life well. They are most often subtle in their communication, but always precise and within that there is compassion and an openness that I find I want to imitate and emulate because I see what good it can achieve in a small sphere. All in all, the cat's meow is fabulous and learning their language has not only inspired me to try to be my full self, but also more.

Chapter 3

My cats make me feel at peace and happy, which is something that most cat lovers who understand the language of their cats can understand. Every household could benefit from having a cat.

Cosmo

Chapter 4:
Nine Lives

The old English myth that a cat has nine lives goes back to the mid-sixteenth century to 1546. Cats were regarded as tenacious because of their careful, suspicious nature and their ability to right themselves mid-fall and land on their feet. They can often survive long falls, though not from the top of skyscrapers as some believe (and have tried to prove with devastating results for the cat). I think of how a cat lives its life every time I open the door to a cat, wondering what he or she might have been up to and what trouble he or she found himself in. How many lives have they already used up?

This phrase also comes from an old Irish legend about witches who turned themselves into cats and back into people again eight times. On August 17th, the ninth time, the witch became a cat for good. August was thought to be a "yowly" time for cats, and could have prompted speculation about witches on the prowl (the source for this information can be found here: http://www.joe-ks.com/phrases/phrasesC.htm). There certainly has been a long time association between the cat and the idea of a witch. In the Middle Ages they were much maligned for the presumed association with witches. Cats were killed off along with their witches and the rat population soared as a result. Cats did not regain popularity until the time of the Crusaders and the Great Plague when they were once again praised for killing rats. Initially, it was believed that cats were transmitting the plague and they were further persecuted and killed on sight. Then, people began to notice that homes where cats were kept unlawfully were able to avoid the plague. It was probably tempting to look upon this as further proof of witchcraft, but homes where there were no cats were overrun with rats and these families were dying of the plague. It was then determined that the rats were the cause of the spread of disease and that cats were the solution to halting its spread. The actual cause of the Black Plague is the bacterium Yersinia pestes, which infects fleas that live on black rats. The black rats spread the disease for the bacterium. I definitely prefer a cat or two to rats and pestilence.

Most of all, the myth that a cat has nine lives implies that through his swiftness and agility, a cat is able to escape life-threatening situations that he finds himself in due to his innate sense of curiosity and adventure. In veterinary medicine there is a saying as well that simply says, "it is difficult to kill a cat." Cats are rigorous. Their bodies want to heal, even when you cannot imagine that they can. Therefore, it is difficult to give up on a cat even in serious illness or injury that at first does not seem to want to respond to treatment. So, a cat might be having problems now, but there are many chances with treatment to do well or succeed. This is true for us humans as well. Taking chances and living fully with intent and persevering in the face of great obstacles can reap great rewards. Looking at our obstacles as lessons the way a cat does can bring acceptance to difficulties. For example, a cat will not get hit by a car more than once, but he may find himself in some new kind of trouble.

Nine Reasons to Adopt an Older Cat

9. **What you see is what you get!**
   Kittens are cute, but they travel at high speed, climbing curtains in a single bound and racing like the Indy 500 across you in the middle of the night. A mature cat knows the word "no," and is more likely to prefer your lap to running laps.

8. **High mileage cats still run great!**
   Pre-owned cats aren't like used cars. They aren't defective or worn out; they may have simply outlived their former owners or become lost and alone. They are more appreciative of our company.

7. **Kittens chew on everything!**
   Adult cats have grown out of this mischievous stage; their taste is now more selective (premium food and kitty treats). They tend to save their energy for more important activities, like chasing a catnip mouse or tormenting the neighbour's dog.

6. **Two well-known clichés about cats are: "Curiosity killed the cat" and "Cats have nine lives."**
   Curiosity usually leads to the loss of about eight of a kitten's nine

lives in the first year. Adult cats are survivors and have experience: "Been there. Done that. Won't go there again!"

5. **Few kittens have mastered the fine art of self-grooming!**
They are just too busy being the mighty adventurer. Adult cats spend up to half their day with personal grooming, wanting to make a good impression when they greet you at the door.

4. **Adult cats require less supervision!**
They sleep more, break fewer lamps and don't try to bite your toes through the blankets in the middle of the night. With an adult cat, you will sleep better, lower your blood pressure, and enjoy their energy and affection.

3. **The truth is: Neither cats nor kittens allow you to teach them anything!**
If you adopt an older cat you avoid the training sessions, failures, and frustrations with the kitchen counter leap, toilet paper chase, and "let's climb up our owner like they are a tree!"

2. **Adult cats don't "litter" as much!**
Kittens build sand castles and even sleep in their litter boxes, and then there is a game called "poo-hockey." People who adopt older cats happily miss this stage of feline fun.

1. **BUT THE MOST IMPORTANT REASON TO ADOPT AN OLDER CAT IS...**
You might be their last chance. Adult cats that end up abandoned, due to no fault of their own, are separated from their loved ones, and are confused and frightened. Many are devastated by their misfortune. Sadly for adult cats, they sit by and watch, as one loving family after another passes them over for a cute kitten. They deserve a second chance for life, love, and a permanent home.

Source for the above information:
www.petpatrol.ca/our-cats/nine-reasons-to-adopt-an-older-cat

This chapter is a glimpse of daily life in a veterinary facility administering to the special needs and concerns of cats as they progress through each stage of life, and as the old cliché says, cats sometimes appear to have nine lives through luck and by good veterinary care. This chapter will tell you how your feline friend can use their "nine lives" to the fullest and live healthy and happy with the one life they do have.

To start with, veterinary care is expensive. That is because it takes a lot of resources to run a veterinary facility at a level of care that is best for your pet. To avoid visits due to illness, it is important to establish a good preventative care program with your veterinarian. I suggest two wellness check-up visits each year to generate a set of baseline medical records and to monitor any changes. Wellness check-ups will note any dental problems, heart murmurs, masses, weight issues, and other general health issues. It is always better to find these things before there are symptoms.

Nutritional requirements will change as your kitten becomes an adult and the adult becomes a senior. One diet does not fit all, so be sure to make adjustments as necessary. I'll talk more about diet later, but I am a big believer that it all starts with nutrition. We can do nothing to change the kitten's rearing or his genetics, but we can certainly impact his health and longevity with good nutrition.

Through studies and in my general discussion with other fellow veterinarians, we have noticed in the last decade or so more and more behavioral and inflammatory problems in cats that visited clinics. There were daily discussions centered on these themes in my practice. The etiology or causes of these problems are complex and multifactorial, but generally a good part of it begins with the beginning of a cat's life. In other words, the etiology or causes are complex and multifactorial—not the symptoms. They are complex too, but my point is that many things come together to form disease. The symptoms rarely start at the beginning of the cat's life, but the etiology quite often does. Sometimes things go wrong even before we meet our cats and this will often map what problems may arise in the future.

The goal, at this point, is not to go into too much detail about the life stages of cats, but to present a basic outline to help readers understand where things can go wrong. In this way you can be alert to possible issues and see your veterinarian if you have any concerns. Prevention, or at least early treatment, is preferable to treating a very sick cat.

Chapter 4

We see different kinds of problems in the kitten, the adult, and the senior cat. The following sections outline the most common feline medical complaints that I saw while in practice. It is my no means complete, but it illuminates how important regular vet care is through all the life stages—not just the kitten stage or the senior stage. Prevention is key. There is no reason why a cat should not reach his late teens or even early twenties, but most don't, so this first section is a real dilemma for some owners.

Cat Years Conversion Chart

Cat Years	Human Years
1	15
2	24
3	28
4	32
5	36
6	40
7	44
8	48
9	52
10	56
11	60
12	64
13	68
14	72
15	76
16	80
17	84
18	88
19	92
20	96
21	100

The oldest cat ever recorded was Creme Puff, who died in 2005 and had attained the age of 38 years and 3 days in human years, which is 169 in cat years. This information and the Cat Years Conversion Chart are from Wikipedia.

Keep Your Cat's Nine Lives Safe: See a Vet

A cat has an innate need to preserve the appearance of wellness, so if your cat is acting sick, he or she has probably been sick for days, maybe longer. You need to seek help right away. Disease in cats is often advanced by the time we see them presented to us at a veterinary hospital. We humans have the tendency to hope for the best—hope it is just a hairball or a sleepy day that has put him off—but it is always safer to check it out if he is not eating well, is sleepier than usual, or is not himself. If he is vomiting or has diarrhea and is not eating, if he is having difficulty in the litterbox, if he is breathing rapidly or panting, then it is an emergency. Your cat's life may depend on it.

For the older cat that is slowly declining and you wonder if you should be seeing a vet, watch for weight loss, increased or decreased appetite, excessive drinking, body posture, or change in routine. If you see a veterinarian at least twice yearly in the senior years, then they will pick up on signs of illness, but things can change quickly in the senior cat. So if you have concern, better to be safe than sorry.

Even cats that do not like going in the car can usually be carefully placed in their carrier by first taking the top off and putting him in bottom first so he doesn't feel he is being pushed into a dark small place. Alternatively, put their bottom into the carrier first. A treat or two can help enormously to keep them feeling positive about the experience. I have seen clients using two laundry baskets held together with bungee cords or boxes taped shut and breathing holes cut into them. I have even seen some cats arrive in pillowcases. It is dangerous to travel with a cat loose in the car. This is less secure than carriers meant for exactly this purpose and I always feel the cat exits a carrier with his head a little higher, his dignity intact.

I encourage kitten owners to take their kittens in the car regularly for short trips to acclimate them to the carrier and the car right from the beginning and continue these frequent little trips in adulthood to keep

Chapter 4

them familiar and happy with the idea. A misting of pheromone spray can help keep him calm. Some prefer to be able to look out the window, while others do better with a blanket over the carrier. Trips to the cottage, to family, to winter sunshine retreats, and visits to the vet become no big deal instead of impossibly stressful if they aren't used to being in their carrier in a moving car. My first long trip with a cat was from Prince Edward Island to Ontario when I was first married and beginning to practice as a veterinarian. It began with her crying a mournful bellowing cry that broke my heart, followed quickly by a huge pile of putrid smelling diarrhea, which she deposited in her carrier and then smeared into her coat. This required a stop and a difficult clean up without the proper means to do it. Many of you will probably understand this if you've travelled with an unwilling cat. The odour lingered for hours and made us all feel sick. She did settle down after five or six hours, but it was a very unpleasant journey for her and for us. Today, there are better options for sedation and motion sickness. Certain nutriceuticals and pheromones can often help. However, we never took this cat with us anywhere again. And when we travelled, she had to stay at the clinic.

I have also worn a cat as a hat and scarf while transporting cats. As a teenager, I had the task of taking a box of barn kittens to have their first vaccines. It was barely minutes before the kittens had wiggled their way out of their box and found my legs, neck, and head. Make sure your cats are well contained. Remember that a loose cat in the car could also become a projectile if you stop suddenly or are in an accident, which will cause injury both to yourself and to your cat. Be safe.

Travel doesn't have to be stressful for cats if we teach them it is not; this can be done by beginning with short trips, treats for positive reinforcement, and lots of praise. A little mist of pheromone spray in the carrier before you go can help a lot. This pheromone became my perfume for many years. It was my trick of the trade. I wear a lovely rose scent now.

Once your cat arrives at your vet's clinic, everyone will admire him so much that he will not mind the trip so much—even if the drive was a bit scary. One of my clients was a trucker, and didn't spend much time at home, so he took his cat on the road with him. The cat became so accustomed to being in a truck that he didn't require any kind of confinement, and just sat beside his owner watching out the window or curled up to take a nap.

Feline practices cater to the cat's needs. They think, Cat, cat, cat, all day long and have fine-tuned treatments. Lately, there seems to be much more focus placed on making practices feline friendly. Cats are not small dogs. They have special considerations where accommodation and treatment are concerned. Examining and restraining a cat for procedures is almost an art form (or a secret language). It isn't really something you can teach; it is something that develops with time. When you work with cats all day long you must develop this skill or perish. You must read the cat's movements and timing and respond like an un-choreographed dance. My technician and I became very good at collecting blood single handedly without the aid of a restrainer. She was much better than me, but I was passable. My first ever attempt at that was with Gus in my own kitchen and afterward I realized that it could work in practice as well. Of course we could not do this with all of them, but there was a level of trust that we had developed with our patients and also a sense of how to hold so it is not too much and not too little, since in most cases, less is more. Good needles, no shaving, and quickness are key. A little growling from the cat is okay, but if their tail is also thrashing, then don't even attempt it.

Regardless of your cat's transportation antics or vet visit demeanour, it is important that your feline friend, just like yourself, gets regular health check visits. Your cat's well-being begins with regular visits to the vet, starting as a kitten and will go a long way to increasing the length and quality of your cat's nine lives.

The Kitten

The first thing you will want to do when you get your new kitten is to have him examined by your veterinarian. This professional will want to talk about many things with you and will want to examine your kitten to make sure everything is how it should be. The veterinarian will ascertain that your kitten is free of fleas, ear mites, and dreadful viruses, and is otherwise healthy and will start the deworming process. They will begin vaccinations as appropriate. The vet will also have discussions with you about nutrition, dental care, and other preventative strategies for your kitten to thrive in his environment.

Chapter 4

Life goes by so quickly when you are a cat. Those kitten days blend into adult days very quickly. At one month old, a kitten weighs about one pound and already eats some solid food and plays. By two months old, most kittens are completely weaned from their mothers and already have doubled their one-month-old weight to around two pounds. It is preferable that a kitten meets his owner by six to eight weeks of age. Hopefully, the kitten has already had other human contact before that initial meeting.

It is essential to gently handle the kitten at this point in its life in order to produce a well-adjusted social cat. If the kitten has had poor rearing for any number of reasons, then socializing with humans becomes even more important. Queen or mother cats who are extremely nervous or stressed by inadequate nutrition or health issues will be less nurturing to their offspring and this may lead to social issues as the kitten grows into adulthood. If a kitten has not had sufficient time with his mother and siblings, then other issues may result as well. You may not notice the social issues right away; after his initial few days in your home your kitten may be a perfectly active friendly kitten with only a small amount of wariness. However, as he approaches social maturity at 18-24 months, his true nature invariably starts to come through.

If the kitten's introduction to humans is delayed, then he may also become a shy or antisocial cat that hides when company comes and doesn't want to sit in your lap. A cat's fear of humans begins at about five weeks of age, so human contact before then is important, but this is too early for the kitten to leave the mother cat. The kitten continues to learn appropriate social skills from both his mother and siblings for weeks yet. The early days with his mother, his siblings, and his owner are important formative days and weeks; if it doesn't happen as it should, then the result can be a skittish or even aggressive cat. If the father cat is an amiable social cat, then his genetics can counteract the female's rearing somewhat. Overall, nurture over nature does have some sway.

Where your kitten comes from is also very important, and is often a factor that remains unknown when we pick them up at the humane society or from a rescue league. Many of them are directly descended from a feral population. While this is admirable, it does mean that these kittens are not very likely to be well adjusted as adults. Their introduction to humans is very likely to have been delayed and there may not have been any handling until after those channels of learning were closed. A stressed and hungry feral queen will have kittens primed by cortisol, which will cause them to have a heightened stress response. Unfortunately, this means that in addition to a tendency to aggression or skittishness, they may also be susceptible to inflammatory problems like interstitial cystitis, which is a bladder problem. This is bad news; but, this is what we expect and this is what I have observed in my feline practice. Certainly, a lot of gentle handling at the right time and the possible genetic contribution from a friendly father can counteract some of it, but rearing and even the environment in utero is important, just as important as it is with humans. Kittens grow and change so quickly that the critical time for learning social skills and appropriate behaviours is brief. If you provide the environment that will help your kitten to feel secure and enriched, then he will probably thrive, but you may only see glimpses of these nervous individuals. We call them cagey because they act like they are caged, since they slink between chairs and around the corner to their bowls or other necessary areas. Learning the origins of your feline can help you to understand their needs, but it is often not possible to come by this information easily.

One problem that I will address in more detail later, is when cats urinate outside of their litter box. This is not usually a problem in kittens, but tends to arise past the time they are litter trained. I was often asked why such and such a problem never happened to cats when the cat owner was young. "I've had cats all my life and never had a cat pee out of the box before." I can't tell you how many times I've been asked that. Well, cats are living different lives now than they did 20 or 30 years ago. More and more cats are indoors only, and often living with multiple cats in the confines of a house where they cannot always escape disapproval or defend their territory. And we are adopting more and more cats from their previously feral situations where they did not have well-developed coping skills for indoor life with humans. This, as well as modern diets are just some of the challenges that today's cats face, but we will discuss more on that later. The modern cat has different challenges than cats who lived decades ago, just as young people growing up today face different issues than people who were their age a decade ago.

Chapter 4

Parasites

It is common to see parasites in kittens, including roundworms, coccidia, giardia, and tapeworms. Kittens may develop severe diarrhea in some cases, and it can take some time to get them back on track. Some kittens may have malaise and a poor appetite, while others may appear normal. Every kitten has the chance of roundworms developing in them. Roundworm eggs are passed through the milk of the mother cat. Cats that get pregnant are cats that go outside, which puts them at risk for parasites as they hunt. Fecal examinations are only 30 to 40 percent accurate in picking up roundworm infestations, so it is better to go ahead and deworm them; worms in young cats can be harmful.

Fleas

Fleas and internal parasites can be very dangerous in the kitten. In fact, they can cause such severe anemia that many kittens will die. I remember a litter of kittens brought to me one time; they were all badly infested with fleas, ear mites, and internal parasites. One of the three had died before reaching the clinic, and another died while I was examining it. I was determined to save the third one. I took this kitten home one evening because he needed to be fed by stomach tube every few hours. Gus, one of my own cats, looked on with concern and interest. I could see I was losing the kitten; he was so lifeless. In a moment of desperation I took a syringe and walked over to my beloved cat. I leaned down and Gus allowed me, without any fuss or restraint, to draw three cc of his precious blood from his jugular vein. Then I was able to find a tiny vein in this tiny almost-corpse of an eight-week-old kitten, and I immediately injected Gus's blood into him. No blood typing, no anticoagulant, and no hope—or so I thought. These were desperate measures, but they worked; the kitten had nothing to lose but life itself. He obviously used one of his nine lives, but he survived and thrived. This kitten went home to his owners and I saw him for years after for his check-ups at the clinic.

Upper Respiratory Infections

Upper respiratory infections are common in kittens, especially if they have spent time in a humane society or other rescue type facility where large numbers of cats of various ages and from various places are housed closely together. These infections may be caused by viruses such as Feline Rhinotracheitis (Feline Herpes Virus-1) or Feline Calicivirus. Alternatively, they may be caused by bacteria such as Bordetella, Chlamydia, or by Mycoplasma. Some infections may become complicated and produce pneumonia in the young kitten and sometimes will be fatal.

Herpes Virus

Feline Rhinotracheitis Virus (Feline Herpes Virus-1) can be fatal to a young kitten, but is more likely to be a nuisance to an older kitten or cat. The virus likes to return; it never really goes away—something like a cold sore in humans. The virus sits in the tissues waiting to show you how nasty it can be if it wants to be. Some cats contract it when they are kittens from other cats and then carry it their entire lives with varying amounts of symptoms. It can cause fever and loss of appetite, so some individuals certainly become very ill with this virus. Imagine a kitten or cat with a chronic green bubble sucking in and out of one or both of his nostrils.

The Herpes Virus has many friends in the bacterial world. It can be a very destructive virus, because it reams the nasal cavity out of its protective mucosal surfaces, which makes it the perfect party venue for various unkind bacteria that produce the giant gobs of yellow or green mucous. The virus often doesn't stop there. It enjoys creating painful ulcers on eyes and even disfigured skin lesions on the face and body. For a kitten or cat whose immune system is impaired in some way this virus can be a misery, and even deadly. To most cats though, it is a nuisance that causes mild to moderate recurring symptoms that often do not invite bacteria and do not require treatment other than symptomatic relief and immune-boosting nutritional support.

When there is a loss of appetite or yellow or green mucous from the nose, then treatment is a must. Cats with stuffy noses will tend to lose their appetite because their sense of smell is linked closely to it. If they cannot smell their food, then they are unlikely to want to eat it. In these cases, softer food may help, or slightly heating up the food may help release more odour from the food. At my practice, we often used saline nose drops to keep secretions moist and to break them up so that the cat was better able to discharge the abundance of mucous. Air vaporizers are also helpful, and taking your cat into the bathroom while you shower can help as well. Your veterinarian will carefully tailor your kitten or cat's treatment to his symptoms. Some need help with hydration, fever, congestion, and pain, while others have different needs. Feline Calicivirus, which can also look similar to Herpes Virus, often causes painful ulcerations within the mouth, which affects appetite. In this case, the cat will appreciate soft food.

I lost a client once after suggesting her kitten likely had Herpes Virus or a Chlamydia infection. She huffed out with her kitten in purse, while saying she had not even had sex yet. I tried to explain, but she couldn't hear me through her indignation. As vets, we don't always remember alternative connotations associated with certain infections.

The Importance of Vaccinations

Vaccination protocols have become a very hot topic among veterinarians and cat owners. Although this is opinion, it may not be every vet's opinion. I know many cat owners who feel they need to protect their cats from vaccinations. But really, the whole point of vaccines is to protect your cat. We are on the same page. Vaccines are important preventative tools and cannot be eliminated from our health care plan. However, I think we need always be mindful of our oath to do no harm.

Initial vaccines in the kitten are essential. They should begin at six or eight weeks of age and be given every three or four weeks until they are 16 weeks of age. The timing of vaccines is important. The first coincides with the time that maternal antibodies obtained through the mother's colostrum begin to wane, and the final dose is given when those maternal antibodies are completely gone. The immune system has a memory response, and so with each booster, the kitten is able to mount a growing response; the final vaccine at 16 weeks will be a fully protective response. This mounting response cannot occur and will not occur if the boosters are drawn out over a longer period of time, if a dose is missed, or if only a single dose is given. The first adult dose given at one year of age is also extremely important because it ensures good immunity. After that, vaccines should be given according to risk or according to the law. Your veterinarian usually assesses this at every annual wellness check.

Adjuvanted vaccine use in cats may contribute to inflammatory issues that we see in many cats due to an overly stimulated immune response. Adjuvants are usually heavy metal additives that enhance the immuno-reactivity of the vaccine, which creates a greater immune response and lasts for longer periods of time. Adjuvanted vaccines have a higher rate of vaccine reactions in cats and they may also cause vaccinal sarcomas, which is a rare but aggressive form of cancer that can develop in predisposed cats. Unadjuvanted vaccines are safer in cats.

Some indoor cats benefit from receiving vaccines on a regular basis. Owners who have indoor cats that live with other cats that go outside for instance, indoor cats that board frequently, or indoor cats that have a tendency to escape should consider regular vaccines for their cats. This is something you will want to discuss with your vet. Outdoor cats of course have greater risk and need more comprehensive protection. Again, your vet can design your cat's vaccine schedule according to their specific risks. Each patient must be reviewed as an individual in order to assess risk and benefit. Risks change with age, health status, and circumstances, so vaccines should be discussed and decided upon at each yearly wellness check.

Rabies is a special case, and every pet anywhere there is a chance of Rabies should have protection from it. This isn't only for the cat's protection but also for your family's protection as well. Rabies is always a deadly disease. See the Rabies Reporter website by Ontario Ministry of Natural Resources for more information on Rabies. In Ontario, it is currently the law to protect all your pets from Rabies by vaccination. Every year in my clinic, there were a few Rabies-suspect cases that we dealt with. It was always on my list of differentials in neurologic cases

Chapter 4

where the pet was not up to date on vaccines. Symptoms can be variable and do not always present as an aggressive foamy-mouthed cat or dog. Unfortunately a diagnosis cannot be made until after death when the brain must be removed and sent for special testing to detect the virus. This takes too much time, so humans involved must be vaccinated in the meantime in case there has been exposure. At that point, it is the health department's concern and it will be their decision as to what measures must be followed.

As veterinarians we are required to report any case that is suspect, and I tended to report even the cases where I had just an uneasy feeling about an unexpected death in a cat that had not been kept current on vaccines. But the responsibility is the pet owner's, not the veterinarian's, and heavy fines can be levied for any pet that is not current on Rabies vaccines. Any time an unvaccinated dog or cat bites a human and breaks the skin, this is a reportable event by law. This requires the pet to be quarantined for a period of time under the direction of the health department and at the expense of the cat's family.

I can remember one lovely couple that frantically came in with their Siamese cat who decided to go out one evening and had encountered a raccoon. They were shocked by this cat's sudden interest in the great outdoors because he had never been interested in going outside before, and on his very first outing, which had not been sanctioned, he met a raccoon. He was fine—not even a hair was out of place. He was a little traumatized by the encounter and close call, but we vaccinated him right away and his owners were furious with themselves for not having done it sooner as had been recommended. They could not imagine beforehand how their lives might encounter Rabies and how it could be a real threat for them.

Another incident that comes to mind is a woman I met at the drug store who was waiting to pick up Rabies vaccines for her family for her doctor to administer, because they had a bat flying around their house the night before and she was concerned that her baby may have been bitten during the night. The bat had been caught and sent for testing but they could not wait for results. They had also taken their pets for vaccines that day.

These kinds of encounters do occur with some frequency. Rabies is a real and present danger for all of us when we have pets. Protect your pets. That will provide some protection for you and your family as well.

Distemper (Feline Panleukopenia)

Thankfully with modern-day vaccine protocols, distemper or cat plague isn't something we see often anymore. The distinctive odour that comes with this particular deadly diarrhea is still in my mental archives, but it has been some time since I've smelled it. The last time I treated a case of distemper was in a young kitten that lived with a sweet teenage girl. The teenage girl took such good care of her kitten and she did everything possible over a period of weeks and months. The kitten survived the distemper, which I thought was a miracle, but the virus left its damage and the cat went on to have chronic gastrointestinal issues that had to be managed. This girl, her mother had confided in me, was very troubled and was suffering from anorexia. The young teenager was so overwrought about her sick pet that her mother worried what would happen if the kitten died. This pet was growing up in such sympathy with the girl that if the girl would not eat, then the cat would not eat. And so the girl ate to ensure that the cat would eat. As an adult, she claims the kitten saved her life—that they had saved each other's lives. I don't think I have observed such a bond between cat and human before or since.

Feline Leukemia Virus (F.E.L.V.)

I learned early that cats often talk us into letting them outside or they just go out defiantly by their own means. I often learned at the first adult visit that a kitten had gotten outside and their status was no longer indoor cat. So, it became standard in my clinic to protect all kittens with Feline Leukemia Virus (F.E.L.V.) vaccine in case they were able to get outdoors. Kittens and young adults are very susceptible to this virus and so if there is any risk at all they need protection. All kittens should be tested for F.E.L.V. It is a simple test that requires just a few drops of blood and takes minutes to perform.

F.E.L.V. is devastating for families. Anyone who has experienced this feline virus in their family kitten or cat will always insist on the vaccine that prevents it, even for indoor cats. Families never want to go through another cat having this virus. Kittens can get it from their mothers and often do. Even before the kitten is born, the virus works its way to produce this awful disease. Like a cruel twist of fate, it waits until we love our kittens and cannot bear to lose them. It seems particularly anxious to disappoint families with young, sensitive children. This virus waits until the kitten has had all its vaccines, has been spayed or neutered, and has begun to be central to life at home, and then it strikes. This virus is especially mean and extremely imaginative in the way it produces disease. It can look like a million different things, but it is always deadly. The first thing I like to do with a kitten is to test it for F.E.L.V., so we know what we've got before attachment is complete, because we become attached to our pets so quickly. If it is a kitten that will go outside, then it needs protection by vaccination so that he does not contract it by a bite wound, repeated or prolonged contact with an infected cat. F.E.L.V. likes to team up with F.I.V. to double the impact in some cats. Many cancers in cats are associated with F.E.L.V.

Feline Immunodeficiency Virus (F.I.V.)

It is a good idea to test for Feline Immunodeficiency Virus (F.I.V.). It is more difficult for a kitten to contract this virus from its mother in utero, as it usually comes from a bite. But it may develop from early contacts. A cat can live with F.I.V if on a good plane of nutrition and if he or she is treated for infections and other disease related to the virus. F.I.V. weakens the immune system and lets other viruses or bacteria do the work for it. Good screening tests exist for this virus, but there isn't a good vaccine at this time that I would recommend except for in communities where the virus is rampant.

Feline Infectious Peritonitis (F.I.P.)

Feline Infectious Peritonitis (F.I.P.) is equally devastating. It is another clever virus. This coronavirus enters through ingestion or possibly through inhalation. It attaches to cells lining the gastrointestinal tract and for most individuals it will cause a mild self-limiting form of diarrhea. The viral particles are shed in the feces and this is why this virus is more common where there are many cats sharing a litter box or grooming each other. In some individuals who likely have predispositions of some kind, this virus can mutate to a deadly F.I.P. form of the coronavirus. Usually we see these kitties in the clinic as older kittens or young adults (less than three years) or older adults (10 years or so), possibly with fever and some recent history of diarrhea. They may also have symptoms related to liver disease, kidney disease, eye disease, anemia, or neurologic disease. They may not be eating well and have experienced some weight loss. In the wet form, the virus attacks vessels and causes them to be weepy. The cat will often present with fluid in its distended abdomen (ascites), or in the chest, or both. A dry form of the virus also exists, which may present with neurologic symptoms or organ failure. They will usually be part of a multi-cat household or from a cattery. Since disease can be subtle at first, it can be difficult to diagnose. Since there is no definitive diagnostic test we can do, sometimes it stays as our top differential for a time before we can confirm it is F.I.P.

For this virus, I think of one little boy's kitten in particular because I doubt there was a cat that was better loved. This cat was just becoming an adult when he began to show signs of sickness. The boy couldn't bear the thought of a loss because he had undergone significant changes in his home life that were devastating enough. His father wished palliative care for as long as we could manage it. As well, the father knew his other cat was at risk. We drained the kitten's belly several times when he was uncomfortable. We also put him on various treatments that helped for some time, but after a few weeks we lost ground and had to put him to sleep. A few months later this cruel virus took the second cat down the same road and the little boy was forced to deal with this loss a second time as well. The cliché of a cat having nine lives does not always win—unfortunately those cats that meet up with deadly viruses only live one life.

Hercules, one of my own kittens, was such a beauty with a long silvery coat with a deep grey undercoat. He had the most beautiful emerald-green eyes. However, Hercules was not strong enough for this virus. At

Chapter 4

about eight months of age, he still only weighed five pounds and extremely thin. He developed neurologic F.I.P. He began to have difficulty jumping onto chairs and climbing stairs. Eventually he began to get stuck in the cat doors and had trouble getting in and out of our litter boxes.

There is currently no screening test for F.I.P for early detection and there is no effective vaccine that will provide protection. There is no treatment other than palliation.

Foreign Bodies

Foreign bodies are common due to how much kittens and young cats love to play. The little barbs on their tongue, which are meant to help with grooming, drinking, and eating prey, also promote foreign body ingestion. For instance, string can get stuck on the barbs, which makes spitting it out difficult and so the kitten will swallow it down. From there the string can cause havoc in the bowels, cinch the kitten's loops of bowels up, and cause painful and dangerous obstruction. Surgery is always required when complete obstruction of the gastrointestinal tract occurs. Further to strings, we have removed ribbons, tinsel, hair scrunchies, threads with needles, dimes and pennies, erasers, pompoms, and parts of things that nobody knows what they are. One of my vet friends removed a diamond ring from a young cat at one time. She said she had never seen a woman so happy to see something on an x-ray as that ring.

Kitten proofing your home is a must. To do so, you need to get down on the floor and look at things from the kitten's perspective in order to spot danger areas. For example, one kitten found a hole in a floorboard and entered some kind of labyrinth within the structure of the home and got lost there for hours. The owner followed his movements above as the kitten meowed and meowed, and try as she may, she could not for some time manage to lure him out. Eventually, the kitten was able to retrace his steps and find his own way out. So, in terms of kitten proofing, once all concerns at floor level, including things that the kitten might chew or that might end up in his mouth are dealt with, then examine the surroundings again at your own height. This includes the counters, tables, possible perches higher up including the mantelpiece covered with precious breakable things, and so on. Try to see things with the curiosity of a kitten.

Poisonings

Poisonings are common in young cats and kittens. They get into stuff and they will try anything. Lilies of various sorts cause fatalities a few times a year. Any part of a lily can shut the kidneys down. It can take a day for the symptoms to show, but by then it's usually too late. It can be a shock to have a lively kitten or cat one day and to discover he is dying the next. I remember one kitten approaching adulthood that managed to spill liquid detergent all over his coat and then he began to try to groom it out. He died even before he got to the clinic; his belly and lungs were full of soap bubbles. Other deadly examples of poisons that young cats and kittens may try include antifreeze, human medications, mouse and rat poisons, and slug bait. Other inadvertent poisonings include medications given to kittens and cats by their families, such as acetaminophen and other NSAIDs, which can be very toxic to cats, and some flea and tick preparations that are available over the counter are also toxic to cats. You can find very comprehensive lists of toxins online.

Cats have very special livers and are unable to produce many of the enzymes required to break down many mainstream medications used in human medicine. This means that any amount of certain human medications could be toxic to your cat. Certain drugs used in veterinary medicine may also be unsafe for cats. Consequently, it is never safe to use human or canine medications on cats without first asking your veterinarian about its safety and proper dosing.

Accidents

Accidents are also common for kittens; strangulation by curtain or blind pull-strings, electrocution by biting through live wires, fractures by getting caught in a door or under a rocking chair, jumping too far or from too a height that's too high, getting hit by a car, or caught in a trap. Sometimes you can't imagine something happening until it happens, for instance, a kitten being shaken by a docile dog that just had his last bit of patience spent. Consequently, kitten proofing is essential.

Chapter 4

Mites and Ticks

Ear mites are common and can cause serious problems for young kittens. Ear mites can cause so much inflammation that the ear canals can close down and permanently impair hearing. Fleas can cause severe anemia and also transmit other diseases, such as tapeworms and blood parasites. There are very safe products on the market today that both treat and prevent flea and mite infestations.

Ticks are becoming more of a concern as numbers are increasing these past many years, which is probably due to climate change. They too can transmit nasty disease causing organisms like those that cause Lymes Disease. Q fever, Ehrlichia, Tularemia, and Ricketsia are other tick borne diseases. All of these are rare, but they may affect your cat or your cat could possibly bring one of these diseases home to you in a tick on their back. It is important to check your kittens and cats regularly for ticks if they go outside. We do not currently have a safe drug to prevent ticks in cats, although there is one that will cause the tick to fall off after a few days. Drugs used for this purpose in dogs are very toxic to cats. There are neat little devices available that remove ticks from your cat's skin—mouthpiece and all. Fortunately, since cats are very good groomers, ticks are generally less of a problem in cats than in dogs. However, the longer a tick is attached, then the more likely it may transmit disease.

To Spay or Not to Spay

It is obvious to me that we must spay and neuter our kittens. The first problem about leaving our cats intact is that they will breed. Cats are induced ovulators, which means it is the act of intercourse that stimulates ovulation. This ensures a very high level of success in breeding and propagation of the species. It also means more than one male may be involved in a single litter. A single female cat will be responsible for hundreds of thousands of cats in her lifetime when you count her own progeny, and the progeny of her progeny on down the many generations. These cats will all need to find good homes and it is not possible for every one of them to find a good home. This is very basic, but there is more to it.

The second problem about leaving our cats intact is that they will want to breed whether we want them to or not. Keeping them indoors may prevent the deed, but will not prevent the instinctual need. Your female will go into heat and pretty much stay there, going in and out of it, through the entire period from January as the days begin to get longer until mid-fall, because they are seasonally polyestrous. She will be anxious, she will be stressed, she will meow incessantly—especially at night, and she will want to get out. This chronic state of PMS will be hard on her and it will be hard on you. Her meow won't be her usual "pet me" or "I'm hungry" meow; it will be a new breed of meow that cannot be ignored—a caterwaul. And if she does go out, she will get pregnant, and so begins the count to hundreds of thousands of progeny. This is why vets do emergency spays all through the months of January through late spring and beyond. The first few heats in a female can be silent. In the early part of the year, nearly every spay we did was while the cat was in heat. During this period, the little uterus, still just barely a straw in diameter, stands up firm and ready for breeding.

Your intact male cat will also want to get out and find that female in heat. It's uncanny how accurate they can be about the location of a female in heat. He will spray, he will make your house smell of his very pungent urine, and he will want to fight. He will drive himself crazy. And when he does go out he will fight and he will be at risk of contracting F.E.L.V. or F.I.V.; he will be at risk of serious bite wound abscesses that will require medical attention. He will have that wanderlust and he will go missing. My cat Tom, the telephone-pole cat, was neutered late because I adopted him as an adult. He never lost that wanderlust and he did go on several vacations on his own. So, it is best to have a male cat neutered by six months of age, but any time is better than no time. Thankfully, it did stop Tom's spraying, which he'd begun as soon as he moved in with us. On the other hand, Beau, our clinic cat, was neutered late and continued to spray with some frequency and to pee in purses and sinks from time to time.

Your female intact cat will have greatly increased risk of mammary tumours later in life. She will be more and more at risk, with each passing heat cycle, for life-threatening pyometra or infection of the uterus. The uterus fills with pus. It is usually a given that pyometra or mammary tumours will occur, and the former will be an emergency surgery that will hopefully save her life. The latter will also mean surgery and may or may not save her life.

I have heard many people complain about weight gain after spaying a cat and yes this can occur. The hormonal changes that occur after ovariohysterectomy or neutering do mean that adjustments need to be made in their feeding. They will require a few less calories and they will still require daily exercise. The daily exercise is often the culprit. Once the cat moves out of kittenhood and into adulthood, that is really when the weight gain occurs. Their energy output decreases but their calories consumed often remains the same.

I can't think of a single reason to keep your pet cat intact. Intact cats do not make good pets. It is the only liberty that I am 100 percent comfortable taking away from them for the greater good of their kind. If we have already taken their freedom from being outdoors away from them, then it is especially cruel to keep them intact when they cannot even breed. If we haven't and they are free to go out and come and go as they please, then they will breed and we will need to accommodate by finding all the multitudes homes, and that cannot be done. If you're not convinced, then visit a few humane societies and rescue leagues to see how they struggle with the numbers.

Many feel that it does not matter whether their own cat is spayed or neutered. I assure you it matters a great deal. You are that important. Your decision counts. For every kitten that your cat births or sires, there is the potential for hundreds of thousands more; most of these cats will never find homes, some will be euthanized, others become part of feral societies where they may never be accepted into the realm, and some will become carriers of disease for all of our pet cats. If they do find homes, then it will only be because the ones waiting in humane societies do not find homes. Their next sneeze will be their last day alive. You may think you can control the outcome of the kittens your own cat produces, but there is no way you can control the hundreds of thousands that each one may also produce through their progenies' progeny. These are the ones that end up dumped in the forest to be somebody's lunch, drowned in a stream, left in a bag by side of the highway, or disposed of at first sign of inconvenience or expense. To avoid these sad situations, spay and neuter your cats. It is the responsible thing to do.

Men often have more difficulty with the neuter conversation. A sort of grimace starts to form on their face that spreads slowly into a look of shock and sometimes anger. One leg lifts and folds over the other in a protective stance as you continue to approach the subject with them. Their expression typically reverses itself from shock to grimace to acceptance when the idea of the female ovariohysterectomy takes place, which I always found interesting and very telling. And, then the finale suggestion comes right on cue: "Well, can't we just spay the female cat and leave the male cat be?" My reply was always the same: "No, not unless you want a very smelly house that you will be embarrassed to bring friends home to, and a cat that constantly gets into fights and needs medical help for abscesses and bad viruses."

THE ADULT

The Outdoor Cat

In the adult cat, if he goes outside then he may need all of his nine lives. These cats come to the clinic due to accidents and injuries involving cars, trees, traps, and other mishaps. It is difficult to imagine the kinds of trouble a cat can get himself into until it happens. We may see him for bite wounds or an abscess, various dread viruses like feline leukemia and feline immunodeficiency virus, or feline infectious peritonitis in the adult cat. Hairballs and gastroenteritis; poisonings are common; broken teeth; blood parasites and intestinal parasites; urinary obstructions and stones; fleas and ear mites; and various common allergic problems including food allergies, flea allergies, and inhalant allergies. We may also see cardiomyopathy in the adult cat.

The Indoor Cat

The indoor adult cat will have some different concerns. For most well-loved single house cats, or even a pair of cats, there should be few if any health concerns if their environmental and nutritional needs are met. Your cat should live a long and healthy life and hopefully only visit a veterinary clinic for a yearly or twice yearly wellness exam. In my practice it was quite common for multiple cats (three or more) to be confined to a small home. Living in confined places shared with other cats can lead to various inflammatory issues due to stress, which will often cause them to void outside the litter box, for example. This particular conversation was my most practiced and least favorite. We humans want what we want. We want our housecats to comply with what we consider to be normal. It is complex and difficult to accept often that it is an environmental concern as much as a medical concern and treatment, therefore has to be multimodal.

Other common reasons for a visit to a veterinary clinic for the indoor adult may be allergies including food allergies, inhalant allergies, flea allergies, obesity, diabetes mellitus, fatty liver disease, cardiomyopathy, asthma, pancreatitis, skin issues, bladder obstructions and stones, dental disease, and various urinary tract complaints. Hairballs and gastroenteritis are also common. Your cat may also have fleas if you live in a flea-burdened region, even if he or she has never stepped foot outside.

It is interesting that allergic disease has taken centre stage in adult cats. I blame modern diets, poorer air quality, and lifestyle. A mild sensitivity or allergy to a food or other substance for instance may result in more severe symptoms in a stressed or poorly nourished individual. Cats and other pets also play an environmental indicator role of which we all should take notice. For instance, the numbers of fleas and ticks on our pets has exploded in recent years. The flea and tick season starts earlier and ends later. And cats are now becoming burdened by heartworm disease, which was virtually unheard of 25 years ago in a Canadian cat. Asthma is also more prevalent. It makes me worry what will be next.

Whether your cat stays in or goes outside, cardiomyopathy is a heartbreaker—quite literally. It is a silent killer. It is often there with no signs at all and can take a cat without warning. Certainly a new heart murmur or abnormal heart rhythm or rate can cause us to be concerned about the heart, and when we hear this on a physical exam it is important to follow up with tests like an ultrasound of the heart in order to confirm whether there is a problem. If there are signs at home then they may include lethargy, exercise intolerance, rapid breathing or panting, lameness, or pain in a limb. Pain in a limb or lameness occurs when a clot has formed secondary to heart disease. When any of these signs are seen at home, they are an emergency and require immediate veterinary care. However, the most common presentation of a cat ailing from heart disease is a cat in full cardiac crisis or sudden death without previous signs or symptoms.

Heart disease in cats generally involves the muscle rather than faulty heart valves. The most common forms of cardiomyopathy (CM) are Hypertrophic cardiomyopathy, Restrictive cardiomyopathy and Dilative cardiomyopathy. Hypertrophic CM is the most common form. In this form, the heart may appear completely normal on radiographs. However, the musculature will be thickened and stiff, which makes the chamber sizes within the heart smaller than normal and the heart muscle unable to contract in a normal way. The heart will be very inefficient and will have difficulty pumping the blood effectively through the vascular system. This will lead to the development of fluid in the lungs or in the abdomen, or both. Dilative CM is the opposite where the heart muscle is much too thin and flimsy; the heart chambers will be enlarged and it too will have difficulty moving blood through the vasculature. The dilative heart will appear enlarged on radiographs. Restrictive CM is an intermediate form.

All cats are susceptible to CM, though there does seem to be a predisposition in Ragdolls, Persians, and Maine Coons. CM does not discriminate; it takes the very young, middle aged, and older cats. Sometimes they are born with it and sometimes it is acquired with age. The Dilative form is less commonly seen now that feline diets are more commonly supplemented with taurine. Treatment of CM depends upon the type of CM diagnosed but will involve medications to improve the contractility of the muscle, to regulate the heart rate, and to help slow progression of the disease. Medications to ease congestion may also be required. Medications to help prevent a blood clot or thrombus from forming may also be indicated.

THE SENIOR

In the senior cat, we begin to see things like chronic renal insufficiency, hyperthyroidism, hypertension, more diabetes mellitus, more obesity, advanced dental disease, pyelonephritis, heart disease, arthritis, constipation, cancers, and senility issues, to name a few. And of course, there seems to be no end of fleas. We see all of these things with frequency, often singly, but more often there will be multiple problems in the older cat. For instance, pyelonephritis, a form of kidney infection, and chronic renal insufficiency frequently come from very poor dental hygiene seen so frequently in cats. Renal disease in general certainly takes centre stage. We have become very good at treating kidney disease and also preventing its progress. This is because veterinarians look for kidney disease early and realize that a urinalysis is liquid gold when it comes to early detection.

All in all, we have explored the three life stages of cats: kitten, adult, and senior. There are many other issues that can affect cats during their stages of life, but I touched on the ones I saw most prevalent in my practice. Although the old cliché, "a cat has nine lives," exists, the reality is that your cat has only one life and it is up to you as his owner to make sure he has everything he needs to make his life healthy and happy. Cats are certainly very independent, or at least they think they are, but they depend on us for everything they need.

Feline Gerontologist

It is possible for a cat to live well into his twenties; in fact, an indoor cat can be expected to live at least until their mid- to late-teens. An 18-year-old cat is about 88 years in human time, and a 21-year-old is a centurion. Becoming a centurion begins with hardy genetics, but otherwise is accomplished by good preventative care; nutrition, dental care, and regular check-ups with your vet. I used to recommend twice yearly appointments for the senior but you should follow whatever guidelines your veterinarian recommends. Wellness testing can find problems early and responding quickly to changes in weight and behaviour will also be key.

Many older cats have multiple age-related issues that concurrently occur. For example, joint pain starts to occur quite young in many cats, and by their teens, most cats could benefit from some changes in the home to make things a little easier for them. For example, some of these changes include having quicker access to lower-sided litter boxes, steps up to favourite chairs, and so on. A subtle change in posture or gait may be the only clue of changes at first. Eating poorly, antisocial behavior, and inactivity could all be related to joint pain. Older cats can get a new lease on life with a little bit of help for their arthritis. There are wonderful diets available for cats with joint pain and sometimes that is all they need. Your vet may also dispense pain medication or certain supplements to help them feel their best in their late years. Joint pain, combined with perhaps some kidney function changes, can lead to very uncomfortable constipation problems in the senior cat. Ask your vet how best to keep your older cat comfortable. Changes in vision and hearing may cause your cat to become more antisocial. A physical exam can determine if polyps or cataracts are a problem. There is little we can do to help them regain age-related hearing or sight loss but we can support them by adjusting their environment as needed.

Without knowing it, I slowly became a feline gerontologist. My first patients were becoming elderly and many of my new clients came to me for the first time with a senior cat. Wellness testing helped determine what changes we needed to make to keep their pets healthy. As kidney disease is very prevalent in senior cats, I became very good at caring for cat's kidneys. When we find early kidney changes through routine wellness testing of urine and blood, there is much we can do with nutritional support alone that can help slow down progression of age-related kidney disease. As mentioned previously, hyperthyroidism, hypertension, diabetes mellitus, various cancers, and heart disease are other frequently seen problems in the senior cat.

However, once things advance, sometimes there is discomfort and pain. I think it is a very wonderful thing to stop pain and give hope for a future of more pain-free days. What I began to feel is that with many of these elderly pets, the treatments given were becoming a misery to some cats with advanced diseases. We could keep those kidneys and other organs going, but at what cost to the cat? And who was this treatment really for?

One of my very last patients that I had before I made the decision to leave my practice was a 12-year-old domestic shorthair (DSH) cat that weighed just three and a half pounds. She had not seen a veterinarian since her initial kitten vaccines and ovariohysterectomy appointments. The family said that she had lived an indoor life and had been healthy until very recently. They had begun to notice some weight loss some time

Chapter 4

ago but did not worry because her appetite was still very good. She drank a lot of water and they found that her litter box was wetter than before. Then she began to vomit periodically and then more often until it was occurring multiple times a day. During this period of time her appetite steadily declined.

When I first saw her she had not eaten in several days and was drinking very little, and that was only from the tip of a syringe. At three and a half pounds, she was skin stretched over bones. She was emaciated, severely dehydrated, and very weak. She had an enormous thyroid nodule, her heart rate was very rapid, her kidneys were small and hard like peanuts, her teeth were rotten and her breath smelt like death. She was most likely suffering from hyperthyroid and she was most definitely in renal failure. The owners refused euthanasia, but also refused all diagnostics, which really would have been pointless anyway. This little cat could barely stand, was too weak to eat, and I did not know what to do. The owner gave me some kind of, "I can't let this cat die," story and I told him I doubted that I could prevent it because she was already in the process of dying. I gave her some subcutaneous fluids and a can of liquid diet with a feeding syringe and she picked up for a day or two; she even ate a little bit on her own, so her owner brought her back for more. I reluctantly gave more and in the process, her paper-thin skin tore—an enormous semicircular area tore away from her tiny emaciated body. I used surgical glue to put her skin back in place and I took her back to the owner and told him I could not in good faith help him anymore, and that it was a very great cruelty to let this poor creature suffer any longer.

Finally, the owner acquiesced and allowed humane euthanasia. This example is perhaps a bit extreme, but it was pivotal to me because it showed me in glaring detail that sometimes we do too much. Just because we can, it does not mean that we should. It reminded me: first, do no harm. I think that if I had seen this cat a year or several earlier, then I may have been able to help her final years be more comfortable. I felt that way many times. Hyperthyroidism is a very treatable disease when diagnosed early. First signs usually include increased drinking and appetite with no weight gained and eventually weight loss. Cat owners usually don't worry at first because the cat continues to eat well, at first. Untreated it puts stress on multiple organs like the heart and kidneys, and results in multiple organ failures. It makes sense to have regular veterinary checkups as your cats become older so your vet can treat treatable diseases and ailments before they get out of hand and while they are still treatable. Waiting until things are dire and then requesting a no-holds approach to treatment won't work in the end. It is too late.

Palliation serves a very good purpose and is a great kindness when the cat is willing and the owner is not ready to say goodbye. It gives us time to adapt to the idea of loss. It can be those final days of caring for a loved one. However, when the cat is not 100 percent on board with medications, hand feedings, and subcutaneous fluid administration or whatever other treatments are recommended, can we still go ahead and feel good about it? Are they worthwhile when we know that these ministrations cannot affect a cure, and can only extend the life for another day or another week or another month for more unwelcome ministrations? Cats do not understand what we are doing. We become a little callous after we see that the medications can be given and we become greedy for that one more day, one more week, or one more month with our beloved pets. I believe that we should let them choose this one last time. Let them choose.

Cat Heaven

Death comes to us all. We start dying as soon as we are born and it is the one certainty of life—it will end. We all must make that journey. I think when pain and suffering can be minimized without excessive unwelcome treatments, then it is okay to let things take their course. When pain and suffering cannot be reasonably managed, then I think that the kindest thing we can do is to provide humane euthanasia.

I know when my time comes, I am hopeful that I will have the opportunity to stay in my home and be cared for there while making my final arrangements and goodbyes. That may mean a dining room converted into a hospice room, a hospital bed, IV catheters, fluids, pain medications, nurses visits, and spending a few more days or weeks with family. Or maybe it will mean a short stay in a palliative care facility. I will not wish to have any kind of resuscitation. I hope to die with minimal inconvenience to others and with dignity. I will want to choose for myself the manner of treatments.

For the elderly or gravely ill pet, we are unable to give this kind of care at home to lessen the labour pains of death. And even if we could, the pet that trusts us to make good choices for him would not understand or welcome it. He would choose something different. You are your cat's whole life and he has lived well; he has no regrets and he says everything he needs to say every single day. He has no further goodbyes to say or arrangements to make. And that is why euthanasia is kinder to them. In the end, you must allow yourselves to grieve.

Clients were often surprised by how much grief they did feel upon the loss of a dear pet. Many have confided feeling silly for feeling so much loss and sadness over the death of a pet. Others confided that they felt more grief when their pet passed than when a human loved one passed and this would sometimes create feelings of guilt as well. Pets bring so much to our lives that it is natural to mourn them deeply. Often our reaction will depend on our stage of life or our situation in life at the time. Often a pet connects us with happier times, a spouse or a parent that has passed, or even the passing of our childhood. When we lose a pet, it sometimes forces us to revisit other losses in our lives and this deepens our grief. Perhaps they have helped us through difficult times. Sometimes we feel a debt to them that we can't possibly repay.

Perhaps your cat has been a source of unconditional love and acceptance that we feel missing elsewhere in our more complex human relationships. They provide that soft touch when we need it. And they are a sounding board or confidante who listens without judgment and responds in purrs and heads pressed into our foreheads. Sometimes we feel guilty for not having done more for such a true and loyal friend, but all we can do is our best and we must be kinder to ourselves. Fighting against these feelings is not helpful. We must allow ourselves to feel it whatever we are feeling. We must allow ourselves to go through the stages of grief and understand that it will pass with time. I think it helps to remember that your cat could not have been better loved or have had a better home. I'm sure he told you that himself every day in his quiet way. Remember that he chooses. And he chose you.

Departing Words

When is it time to say goodbye to a beloved elderly pet? This is a very good question and the answer is different for every cat and every family. I like to ask clients to write a list of all the things that make their cat who they are, the things that epitomize them. The following questions will help you this about what characterizes your cat:

Do they wait for you at the door when you come home?
Do they sit in the deep red chair beside you when you read the paper?
Do they sleep at the foot of your bed?
Do they nip your ankles at mealtime?
Do they chatter at the birds outside?

Write a list of a dozen things that describe your cat's personality and daily life. Once he stops doing most of these things, and especially if they are not replaced by new daily activities, then his quality of life is deteriorating and he is waiting for the end. Also consider his appetite, posture, whether he is grooming, his ability to use resources like the litter box, and any abnormal things he may be doing, like vomiting or new postures that suggest he may be having difficulty breathing. Cats are tenacious. It takes time, sometimes a lot of time, and often a lot of suffering before the end arrives.

Sometimes the euthanasia talk comes too early in a cat's life when there is grave illness and little promise for recovery. This is especially hard to accept. Sometimes it comes early because we have not planned for the possibility of illness in our pets. Pet owners need a contingency plan. I have seen well-meaning families come in with cats that have disease or illness that is not life threatening, but it nonetheless takes their life due to late treatment or lack of funds.

My feelings about euthanasia have changed over the years. I admit that I have some confusion about it. As I have become older myself and as I have performed more euthanasias than I care to count, each one seems like a new kind of countdown. I did not consider when opening my clinic that a clinic also has a life cycle. After 17 years it did seem euthanasias were becoming more and more routine as my first patients were nearing the end of their lives. Many others however, I have felt forced to do because I wasn't able to promise a quick fix and needed to put a poor creature out of his misery. I will never perform euthanasia under these circumstances again—somebody else will have to. A no-kill policy does not always work.

I know the blame does not rest with my fellow veterinarians or I as we perform these deeds; it is not a result of my neglect or cruelty. I am the only one in the examination room that can take the pain away, but the decision to wait or not to wait—to treat or not to treat—is not mine. And yet, I am the one with the needle in my hand. I am the one who has to give it. Even when given with love and good intent, the euthanasias began to feel that those injections were an aggression towards myself. The irresponsibility of another forced me to put my own morals aside and to give a final solution to suffering instead of using my skills to heal. There is a feeling that overcame me when I pushed the contents of the syringe into the cats' veins. I felt a swoosh, and with it an equal and opposite simultaneous swoosh of the pet's life force reverberating through me. I still think of that double swoosh, and I can still feel it. I am the one who also feels that she must answer to God for it. I have so much regret.

For most euthanasias, I felt that I was giving the gift of painlessness—the gift of an easy loving release into death—which was an escape from the labour pains of certain death. But for many, where I knew help came too late or without proper intent, I felt a piece of my own soul escape with the soul of the dear creature through my fingertips. The personal cost was too high for me to continue.

The idea of nine lives is rather lovely. A cat uses most of his or her lives in his first year or two, after which he has learned so many lessons; he is then very likely to live a long and prosperous life. I think of a cat's extreme physicality as a metaphor for the human psyche. Physically speaking, they push things to the edge of what is safe or their comfort zone, but they learn from that. They are not exactly reckless, because they usually know their limits and they push it to the very edge of that comfort zone. It usually works out for them. Humans perhaps could learn from that nimbleness of their spirit, that confidence that a thing can be done and just to go for it, and that we should push ourselves to the very limit of our own comfort zones. Like cats do, we should push our comfort zones to fully and authentically express ourselves in order to be our full selves, but not to do reckless and dangerous things that may cause injury. Like me writing this book for example. It is a definite stretch to the edge of my comfort. As your cat's closest friend and protector, you will watch him change from a kitten to an adult and then hopefully to a wise senior cat who will teach you some precious life lessons of how to live a life with nine lives.

Chapter 5:
On the Wrong Side of Every Door is a Cat: The Indoor Cat

"On the wrong side of every door is a cat," is a cliché that reiterates the "grass is always greener" concept. Anyone who lives with a cat knows that kitties always want to be on the side of the door where they aren't currently standing. The origin of this phrase may be T.S. Eliot's 1939 poem, "Rum Tum Tugger" (from Old Possum's Book of Practical Cats, on which the musical CATS was based). The character Rum Tum Tugger is never satisfied and is always on the wrong side of every door.

A Purpose-Driven Life: Is The Role of Companion Enough For the Indoor Cat?

Every creature under the sun requires a purpose; otherwise, their life force shrinks—they do not reach their potential or their true selves. This is as true for a cat as it is for a human, especially when the cat lives indoors. It occurs to me often, when I see a cat outdoors meandering about and inspecting his surroundings, that the word housecat is an oxymoron.

It has been my observation over many years, though I don't think it was immediately obvious to me, that many cats do not find their purpose in life. Perhaps they cannot find it within the indoor world we have constructed for them. But, like many cat owners, I have indoor cats too. I am afraid to lose them to the big bad world out there. Of course, many cats acclimate to the indoor lifestyle, because they have never known an outdoor lifestyle. However, the genetics of the species has not evolved quickly enough for cats to tune down their genetically coded traits in order to fit into the confines of an indoor environment the size of a house. Are we doing our cats any favours by keeping them from scraping their knees and dealing with bullies? Are we catastrophizing our cat's lives and eliminating certain risks in exchange for more palatable risks?

Now, living house bound certainly can suit a certain population.

Persians, Ragdolls, and Himalayans, to name a few, belong on a velvet cushion looking out a window observing birds and life on the other side. It suits them fine; it gives them time to contemplate the meaning of life like a Buddha. But what about the other breeds of cats?

A cat's purpose may be as simple as being a companion to an elderly widow, or a person suffering from depression and loneliness, or someone who is sick and bedridden. This is a very worthy occupation and a cat will do a marvelous job at keeping people company. A cat understands the companion role and feels the purpose. The human being and the cat look after each other. I sense the human of the pair learns to take better care of himself in the process. I've heard many senior clients claim their cats are so engaged that they soon expect them to be answering the phone or handing them tools as they sit and watch them work at a project. Nevertheless, the other question remains: Is a cat happy and fulfilled living indoors?

Tabitha: A Story of an Indoor Cat Who Kept a Senior Company

One story about a cat named Tabitha highlights how the life of a senior can be enriched by the attentions of a cat. An adorable eccentric senior used to talk about dust or silt overcoming her home, piped in by the pimps in the neighbourhood, because she had figured them out—she knew they were trafficking prostitutes and drugs. I so much wanted to understand what this dust or silt was and if it really came from the chimney, but this lovely lady had lost the ability to use all her words at times, and I could see and understand her frustration with that. However, she was still fully present and she smiled, which let me know there was no need for concern.

She felt that Tabitha kept her safe from her well-loved neighbourhood that had taken a turn towards the darker side. She did not seem afraid, but not hopeful either. In essence, Tabitha gave her a reason to trudge on. We had regular contact with this owner's son and he was aware of the silt and dust piped in by the pimps, but we did worry about the senior. She usually called the clinic weekly with updates on Tabitha's health and in that way we kept an eye on her. The owner insisted occasionally that

Chapter 5

Tabitha needed time at our facility, like a spa vacation, and she would have a taxi drop her off for a week of repose. I would argue with her that it was stressful for her cat. She would argue that it was recuperative for her pet and that she had discussed it at length with Tabitha. Who was I to argue with her?

The owner felt Tabitha needed a break from the dust that was coming down the chimney and she insisted that Tabitha always came home a new cat after her visit to my office. I would try to explain to the owner that she came home a new cat because she was so relieved to be home. However, the owner countered that I couldn't possibly understand how changed her pet was when she came home—she said it was like she had a renewed interest in life. Tabitha would be playful and more social after spending a week with us. I'm still not sure what to believe. Did Tabitha really enjoy her time with us? She mostly just seemed uninterested in our affections, but maybe she was interested in the activity around her. She and Beau certainly had developed a friendship of sorts.

I decided to believe in the trust the senior had placed in us. I accepted her pleas for us to accommodate Tabitha as a regular visitor and with her consent, Tabitha enjoyed all the privileges of our clinic cats—out and about all day and confined at night. On one of these visits I felt the timing was very fortuitous indeed. Tabitha's owner had reported that a squirrel had come down recently with the dust and attacked Tabitha and she thought she might be wounded. There was no wound, but Tabitha did have a high fever and a very severe kidney infection. She stayed with us for four weeks that time while she got back on her feet, because her owner was too nervous to bring her home.

Tabitha had all the privileges of a pampered cat. According to her owner, she slept on lace undergarments by her own instructions and spoke perfect English. Her owner would tell me about many of their conversations and Tabitha's opinions on many things. She was a very wise and forward thinking cat, just like her owner. She only ever spoke in meow to me, but I didn't know her as well as her owner knew her. I never knew for sure if she was happy as an indoor cat or longed to be on the other side of the door, out in the world.

A Cat Tail That Reflects the Plight of an Indoor Cat

As I write, I am aware of a drama occurring outside my kitchen window. My cats are also extremely interested. It involves a pair of house sparrows who are preparing their nest in the beautiful ranch style birdhouse high on a pole in the yard just feet away. The cats are watching intently with their tails flicking and their occasional chatter tells me to look for myself. The sparrows play out this same drama year after year. There may even be a clutch of eggs in the birdhouse already. The sparrows go in and out with bits of grass and other materials that they scavenge from the yard. They also go in and out and remove old materials that are unsuitable.

Once their eggs are laid, a starling becomes interested in the sparrow's birdhouse. I have chased him away already a dozen times today. The starling will eventually succeed in stealing an egg or two and the female house sparrow will have to lay others to complete her clutch because this is what they do. On occasion, they have succeeded in hatching and fledging a clutch of baby sparrows and gone on to lay a second set of eggs. But the odds are against them. In about a month, a pair of house wrens that are a fair bit smaller than a sparrow will chase them away and take over the house. The starlings won't even dare to check in while the wrens are in residence.

The sparrows can't help themselves. It is instinctual for them to come back to their house and try again, just as it is instinctual for the starling and the wren to play out their parts in this drama and for the cats to watch it all from their window. It is rather frustrating to watch. But you have to admire their perseverance at least. I could move the birdhouse, or change the birdhouse to one with a smaller door hole so that the sparrows would not consider it for their home any longer. It is that occasional clutch of chicks that they do hatch and fledge that keeps me from doing it. It is so fun to watch them feed them and to see them take their first flights. In the great outdoors there are millions of other similar dramas that are played out in addition to the ones in our neighbourhood alone.

The house sparrows will attempt to oust the wrens for a day or two and then they will disappear for a day or two while they go house shopping.

Eventually they return and set up house in the cedar kitty corner to their preferred home and they will be successful there.

The entire drama reminds me of our indoor cat's plights. A cat is not a house sparrow. Its cognitive skills are far more advanced. However, they respond instinctually to their environment and we cannot place blame. Out of doors, there are options. There are sometimes few options indoors that harmonize well with a human's sense of what is right. We can make things easier for them to cope and when there is a bully in the house, more protection must be afforded to the bullied cat, because they cannot move house the way the house sparrow can. Resources have to be made readily and safely available to each individual.

There is a second birdhouse in our front yard amidst a few tall fir trees and shade garden with beautiful ferns, hostas, and Solomon's seal. This little ranch bird home produces a few sets of fledglings every spring and summer. The pair of sparrows can be seen sitting on their front porch sometimes in the summer evenings, with not a care in the world. Their environment is protected and they thrive there.

Cat Obesity: Health Concerns

One of the cruelest things we can do to our cats is to allow them to become overweight. This problem deserves special notice because it is so prevalent. It is a difficult problem to manage. It is prevalent for a number of reasons, all of which are modern.

The first reason is that our cats are staying in more now and not getting the exercise they need to stay trim. Add to that the eventual apathy that often occurs with indoor living—inactivity begets more inactivity. And then the extra weight makes activity more difficult. So, a vicious cycle of weight gain and further inactivity begins. No different than we humans.

The second reason is the modern cat diet and method of feeding, which is often free choice these days. A cat's satiety centre is triggered to tell him he has had enough to eat by amino acids, which make up proteins. In humans this is actually glucose, the shortest unit of carbohydrate, which triggers our satiety centre and tells us to stop, or stomach distention for us who eat too quickly—whichever comes first. Today's mainstream diets often have such poor quality indigestible proteins in them that the cat has to keep eating in order to try to reach that satiety state. In the meantime, along with the poor quality proteins, he is ingesting a large quantity of fat, carbohydrates, and calories, especially in many dry food formulations, which the cat must then deal with in some way. Cats do not require a large amount of carbohydrate in their diets. They are able to make carbohydrates from proteins by gluconeogenesis. Large amounts of carbohydrates lead to insulin resistance and fatty deposition. The fat itself acts in a way to promote further fatty deposition. So another vicious cycle occurs, which favours more weight gain. At this point, a cat does not even need to overeat to continue regular weight gain. In fact, less and less food will be required to maintain an obese cat's weight. That is why it is so frustrating and difficult for our cats to lose weight. We cut them back and cut them back from what seems like not very much food to begin with and they still don't lose weight.

It is so frustrating that many of us just give up. A fat cat is cute, right? But there are a few consequences that are a given if we allow obesity to occur. The first has to do with hygiene. He will no longer be able to adequately groom himself in certain areas, and this will lead to mats along his flanks, rump, and hind limbs that will become painful. He will be soiled at his rump and this will lead to urinary tract infections, anal gland impactions, and possibly abscesses. The second consequence is that he will be inactive and this will lead to muscle wasting, further inactivity, and mobility issues, including more stress on joints and arthritis. He will have trouble maneuvering in his litter box, he may miss the edge, or he may not even feel comfortable getting into it anymore. If his litter box is too far away, he may just choose his own spot. He will not be able to jump into his favourite perches, and this will lead to further boredom, which could possibly mean more eating, but it will certainly affect his quality of life. He may suffer from constipation as well, which will increase his discomfort. The third thing is that he will develop one or more of the following health issues directly related to his obesity: diabetes mellitus, heart disease, fatty liver disease, pancreatitis, arthritis, and many forms of neoplasia or cancer. With diabetes, it is almost a certainty that it will happen—the real question is when it will happen.

Treatment for obesity is multipronged and your veterinarian should regularly monitor it to make sure weight loss is achieved safely. It involves regular daily exercise. Even a little walk on a leash in the back yard twice a day will help immensely. The tiniest bit of weight loss will make the next little bit of weight loss that much more possible. The tiniest bit of muscle tone gained will also make it more and more possible. The tiniest bit of interest found on these walks also feeds back to further interest developing in the physical world.

It also involves a good quality soft and measured reducing diet high in good quality proteins preferably divided into several meals per day. Eliminating treats or most treats will be necessary.

I had a client that insisted he did not overfeed his cat, giving exactly the amount we calculated for weight loss and not a crunch more. But, the cat continued to gain weight with each visit. At one point I asked if he'd stopped giving treats as we had discussed and he said, "Well, no, he insists on his treats."

"Well, how many treats does he get?' I asked.

"Probably about a cup a day," he said. A cup a day!

If you feed your overweight cat a dry food, then it can be a good idea to turn a meal into a game by placing it in a rolling food dispensing toy. Treatment for obesity involves regular weight checks under veterinary supervision to ensure that weight loss is occurring in a safe way. But the best treatment by far is prevention and that can only be achieved by ensuring your cat is getting adequate nutrition especially good quality proteins, counting calories, and keeping your cat active. Most overweight cats are indoor cats.

The Indoor Cat: Consequences and Real Risks, Including Euthanasia

A large portion of what we treat in veterinary medicine has to do with indoor cat status. There are many risks of letting a cat go outside; I do not wish to minimize that in any way. Going outdoors can lead to many serious problems, including death from any number of encounters; cars, predators, dread viruses, poisonings, or cruelty just to name a few. Further, outdoor cats may also fall victim to bite wounds from other creatures, usually other cats that may transmit F.E.L.V., F.I.V., or even rabies, abscesses, parasites, or become pregnant. Sometimes we just never see them again and have no idea what has become of them. This is particularly difficult to accept; it is heartbreaking, especially for children to cope with a missing beloved pet.

However, it is important to understand that an indoors-only cat carries its own set of very real risks that can directly result in death, often from euthanasia. In fact, behavioural problems in veterinary medicine are some of the mostly deadly problems that we see. And usually these problems result as a normal consequence of normal feline behaviour. The most common feline behaviours that can get them ousted from their homes, possibly euthanized, or surrendered to humane societies include voiding out of the box, spraying, aggression towards other pets or humans in the home, and destructive behaviours such as scratching the furniture, rugs, or walls.

Outdoors, cats have nine lives (or chances) so the saying goes; indoors, they often just have one chance. I am not exactly advocating that you let your cat out. Just that we cannot make a decision to keep a cat indoors, shrinking their life down to the size of a house, without understanding that there are consequences to that. Imagine if we kept our kids indoors all the time and never allowed them out except in the car in a crate twice a year to see the doctor. You'd expect some problems wouldn't you? I know in some communities there are bylaws that demand that our cats stay indoors, and I think this is a shame, especially if they are inclined to go outside and we accept those risks. It seems backwards. Natural ways of living, using nature to our advantage, is not always indecent. Of course, bylaws need to be respected and followed, but there are still ways to allow safe access outside in these cases; for example, you can use outdoor enclosures. As with all things, we must try to look at it from every perspective and one of the things these bylaws try to achieve is to limit the loss of our domesticated cats into feral societies, and therefore reduce the numbers of cats living this way. It can be a miserable hard life for them.

There is something organic about sunlight touching human skin and absorbing into cat coats. There is something vital that happens when human toes and cat paws walk in the dirt that is essential to our life force. I truly believe this. Clinton Ober, Stephen Sinatra M.D., and Martin Zucker, in their book *Earthing: The most important health discovery ever?* explain how going barefoot outdoors a part of each day actually discharges positive charge that builds up in our bodies and contributes or possibly even causes inflammatory processes within us. I immediately thought of all of our indoor cats when I read this book and all the inflammatory problems we see in this species. I garden in my bare feet all summer long and I do find that it enhances my mood. It feels good. It feels natural. It is an excellent cure for a headache at a minimum. We cannot see what other effects it may be having within us, but I am a believer. Inflammatory ailments can be so subtle. For example, I hear from my own family doctor that Vitamin B12 deficiency is common in humans (as it is in cats) and I wonder how often it is related to undiagnosed inflammation. It often seems that when this deficiency is found in humans, the cause is not always sought. Symptoms of the deficiency itself (fatigue or apathy, for example) may be its only sign. The deficiency demands treatment but what is the cause of it? In cats, B12 deficiency occurs often from inflammation in the ileum, the part of the intestinal tract where Vitamin B12 is absorbed or in the pancreas, because the pancreas is responsible for producing an intrinsic factor required for Vitamin B12 absorption. Other forms of inflammatory disease are common among humans and cats as well. And we've all heard about Seasonal Affective Disorder (S.A.D.), which so many of us, or maybe all of us, suffer from during the wintertime due to shorter day length and quite possibly less time spent out of doors. Imagine if we never went out, got our feet dirty, or had the sun on our faces at all. There are always multiple factors that combine to cause disease or disorders starting with a genetic predisposition. S.A.D. is not yet fully understood in humans but it does appear that serotonin levels are lower during winter months when hours of sunshine are lowest and this is believed to be a factor in producing symptoms of S.A.D.

I have come to think of many cats with behavioural issues as though they have chronic pan-Seasonal Affective Disorder. Their lives are mostly unseasonal when they are confined indoors without benefit of direct sunlight. We try to fix them with antidepressants like Selective Serotonin Reuptake Inhibitors, Tricyclic Antidepressants, and Monoamine Oxidase Inhibitors that increase the serotonin levels at the synapse level in the brain. Serotonin is a neurotransmitter that promotes a happy relaxed mood and that is why these drugs are often prescribed. These drugs work very well for many individuals. Serotonin is produced naturally in the gastrointestinal tract and in the brain from tryptophan, which is an amino acid found in proteins. Vitamin D (sunshine), exercise, and good quality nutritional proteins all work towards promoting the production of serotonin. These are elements that are often missing in the modern cat's lifestyle.

Taking a pill for what should occur naturally for most individuals given the right environment and foodstuffs feels wrong. There are some lovely nutriceuticals (dietary supplements), however, one in particular I chose to use that contains a casein (a protein) derivative that has a nice civilizing happy-go-lucky effect similar to serotonin in cats with behavioral problems. Over all, we can give any number of things to enhance a process that would occur naturally in most cases in a natural environment. But, should we? Should a cat be confined indoors? Is this natural and the best environment for the cat? There were lots of days at the clinic where I felt like all I needed was a prescription pad in my pocket inked with the words Let Kitty Out QID (four times daily). But it is not that simple because other days what I needed was a prescription pad in my other pocket inked with the words House Arrest: Keep Kitty In. So, there are risks to both. We need to find a balance. If your cat suffers from behavioural issues or inflammatory issues possibly related to stress or his indoor lifestyle, then talk to your veterinarian about best treatment and environmental enrichment options available for you.

A friend of mine recently contacted me on Facebook with an interesting post about a famous study by Dr. Masura Emoto, which demonstrates that our thoughts have a direct effect on the structure of water molecules within ice crystals ("Explore," 2006, No. 5, Vol. 2). The double blind study shows pictures of ice crystals that were formed in the presence of people thinking about specific things, like anger, jealousy, hatred, love, forgiveness, gratitude, or prayer for example. The crystals formed were beautiful and symmetrical for all beautiful thoughts, and they were very ugly and distorted for the ugly thoughts. This was just further evidence to me that we are naturally energetic beings, and it matters greatly what we put out there through our consciousness into the universe. Given that our brains are 70% water, it makes me wonder what sort of effects our negative energies have on the structure of our brains, bodies, and even on the world at large. If this study is valid (all scientists will be skeptical because it is our nature), then it raises larger questions about the power of our collective thoughts and how it may affect other parts of our physical world and our bodies. There are many in the scientific community that want to believe in these energies but feel that the scientific methods and approach to the study were not strict enough to draw conclusions. So, we wait until these results can be reproduced under more rigid laboratory test designs and conditions. But, the idea is intriguing and still has merit. Once again, when I think of this, I am reminded of the indoor cat's plight that has inflammatory and behavioural issues that plague him. His life is often in conflict with his nature; can this affect the structure of his water molecules within the body and brain? We will likely understand more and more about how our own minds and thought processes affect our general health in the future.

In the beginning, when cats were first domesticated, it was the cat's idea. They essentially underwent self-domestication. Some argue that even now they are not fully domesticated because they can revert to total self-sufficiency at any time. They are the quintessential opportunists. They recognize an opportunity and they seize it.

When humanity first started forming communities in the cradle of civilization, when they gave up the nomad hunter gatherer lifestyle, settled down to farm, and domesticated animals five to eight thousand years ago or longer (depending on who you ask—a cat or a human), the cats began to be interested in human activity. With the settled life, there was plenty of grain to feed the humans, cows, and goats, and so the mice and the rats came, and after them, came the cats. Humans readily accepted the cats because they realized that they could decimate a population of mice and rats rather quickly. I imagine in the beginning it was a loose association, that each tolerated the other, but then as cats developed some trust and humans developed more reverence, they were invited into the homes of humans to take care of rodents there. They had a job but they were still free to come and go as they pleased so they could still be in society with other cats and breed, mark their territories, and make their rounds patrolling their territories to make sure their status was secure, which was their true purpose.

In pursuit of keeping our cats safe, we have unwittingly taken away their higher purpose and their small and big pleasures: walking through the grass, observing from under a bush or up in a tree, marking their territories by various means, patrolling their territory, scratching tree bark and other natural surfaces to condition their nails, hunting, choosing their own areas for elimination, following a bug or butterfly, and forming societies with other cats of their own choosing.

Yes, we have saved them from the cold, the need to eat mice, and from all the other perils of outdoor life. But, would they have chosen it themselves? Would they have even made the trek from warmer climes if we had not taken them along with us to manage rodents upon ocean vessels so many years ago? Many of them probably wouldn't have, preferring warmer temperatures year round. But as we have seen, cats are very adaptable.

Many would argue that their cat would not go outside if you left the door open all day long because they are so disinterested in having anything to do with the outside world. In fact, many cats that are used to the indoor life don't even want to get their feet dirty. Many wonderful clients have shared this experience with me. However, is it disinterest, or is it fear? Or maybe it is apathy, and where does that come from? It is a bit like believing that all free-range chickens actually go outside. In many cases, their introduction to a small flapped door in the huge barn to the outdoors comes too late. Perhaps, the chickens have too much fear or maybe just disinclination to go through the small flapped door, to face the bright sun, the fresh air, to experience a change from routine. The claim can be made that they are free range but they are not truly free range. This doesn't mean you should force your cat to go outside; that would be cruel. However, your indoor cat (who chooses to stay indoors, even when you offer him an open door to the outside) may fall into the same statistics of cats that are confined and want to go out. It is the state of being indoors that is the risk, not the desire to go out. Nevertheless, risk for problems, such as behavioural problems, to occur increases manifold when there is crowding, or too many individuals in a given space.

Crowding can be another major problem for cats. A dozen cats may live together in perfect harmony confined in a small home without mishap, but that is rare. With multiple cats living in a closed space, the number of interactions per day between individuals can be very high, often to a point where coping can be difficult for some individual cats. Even in the wild there are stocking densities that occur quite naturally and that work, and once imbalance is met, something happens to settle it again. For instance, when the deer population explodes, coyotes move in, and that is just a small filament of the web of connectivity in nature. It is not just about availability of food though. Have you ever shared an apartment with multiple classmates? It can work very well, but when things get too cozy, we can get up and leave. Think also about pigs, intensely farmed in today's modern factory style with minimal space and no natural light. As they are fattened in tiny spaces, piglets begin to savagely chew each other's ears, toes, and tails. They are so out of their minds because of their abnormal environment and limited space that they turn to cannibalism, but not from hunger of the stomach—it is hunger of the soul. This is an extreme example of how limited space can affect domesticated animals. Pigs are very intelligent animals and this kind of suffering from the day they are born until they die isn't worth the bacon. I have to think we will answer for it eventually. The same might be true for keeping cats inside. In fact, numbers matter.

Fleas Indoors

A popular joke in my family is, "Have you treated any fleas lately?" As though fleas were patients of mine.

I spent a very large part of my days, spring through fall, talking about fleas and treating fleas. I'm not fond of fleas, but you have to respect their persistence. Their life cycle and their tenacity cannot be matched. They are here for the duration, although they are rather nasty because they carry some other nastiness with them.

Your indoor cat can get fleas. They will often be on the wrong side of the door. It will not usually happen in the early months of the flea season. Usually they will make their appearance as their numbers peak in late summer and fall. Some of you have had the occasional mouse in the house I'm sure. I know I have. A flea is every bit as able to find his way in and then some. Often he'll come in on the back of your cat, but there are other ways for him to find his way in. It's not that difficult for a tiny little creature like a flea that is motivated to find a warm place to winter. They flee indoors and stay there if they find a nice warm cat or dog to feed off. This is more likely to occur if you have the kind of yard that fleas like. There will be good years for fleas and bad years. But overall, their glory days are just beginning. The flea season in Ontario at least is starting earlier and is now extending further and further into the fall until we've had a few good freezes; the extended season compliments of climate change.

A quick review of the flea life cycle may help for you to picture it better. Each adult female flea lays hundreds of eggs after a blood meal. She lays them in your cat's coat and from there, these slippery little eggs disperse about your house every time your cat gallops like thunder down the hallway or jumps up on his favourite chair. Those flea eggs are perfectly designed to be very nearly invisible and readily dispersed. They will develop from the egg, which could be vacuumed up quite easily, into a sticky pupae that cannot. They hatch into the baby flea within 21 days and the cycle begins again. In this way, flea numbers balloon quickly from a single flea to thousands within a few months. They also have the ability to suspend their development, to stop development if their environment goes quiet. This is why when we move into a new house that has been vacant for a few months, or into a summer cottage that has been empty all winter, we can experience what seems like a sudden explosive flea infestation after a few weeks. They begin again to develop from their arrested development when the vibrations and commotion of life around begin again. You have to admire a flea for his stick-to-it attitude and ingenious life cycle.

It is never a surprise to me when an indoor cat comes into the clinic with fleas. I expect it. However, it is usually a shock for their families. You can sometimes see an offensive stance building and then the words, "He must have caught it here," and even when you show them the little crumbly bits of flea feces in clumps within the coat that have been deposited over weeks, it is still difficult to fathom. It happens; it's okay. Like lice in kids, it is not a reflection of their parents, home, or hygiene. It is just life. It happens.

Chapter 5

Fleas can be elusive when they are in small numbers. The best way to find them is to look for evidence of them. A flea comb is my favorite tool in the exam room for this. A flea comb has tight metal prongs for combing and it is good for catching fleas but even better for finding evidence of fleas. Flea combs are also my favorite tool for combing cats to keep hairballs under control. If the number of fleas is small, then you won't catch one. But you will find their feces. Comb around the neck, down the spine, and along the tail head, and you will find evidence if they are there. The flea feces are crumbly and black. Some bits of it will be coiled. If you put it on a white tissue and add a drop of water, then it will turn a rusty red color. It is digested blood. The eggs may also be visible when you use a comb. They will fall to the table and look like little grains of salt. The two combined can look like salt and pepper. It was common to get a few phone calls every summer from someone who had just bathed their cat and called to report that he was bleeding, but they could not find the source. Basically, it was the flea dirt that they were seeing dissolved in the soapy water and running off their cat like blood.

Some cats, especially outdoor cats, often don't care about a few fleas and won't appear to notice too much. The indoor cat that is less familiar with fleas and their tickling may appear very agitated and frantic by a single flea, and maybe as shocked as you. Fleas favour certain individuals over others so in a household of cats, one or two for example, may seem to carry a heavier population of fleas, while others may not appear to have any. It is important that all pets in a household be treated for fleas and maintained on a preventative treatment; otherwise those less affected become sources for reinfestation. We should care about these fleas, whether the cats go out or not. I know many feel it is inevitable for outdoor cats to get fleas but it makes a lot of sense to try to prevent a major home invasion. And it makes even more sense to prevent the terrible things they can transmit to your pet including tapeworm infestations, life-threatening blood parasites, flea allergy dermatitis and hotspots, and even anemia in young kittens and elderly pets.

Flea treatment and prevention is easy with modern veterinary products applied topically monthly. Often additional environmental products are not even required. Please speak to your veterinarian before choosing flea products. It's important. There are many over the counter preparations that hopefully will be illegal to sell someday soon. They are not safe. In the olden days all we had to work with was organophosphates in form of powders, sprays, and shampoos. We knew these products were not ideal for cats but it is all we had, so we used them judiciously. Now they are sold in concentrated form as topicals, which makes them even more dangerous. They are cheap to produce and less expensive to purchase than a veterinary product, but are expensive enough especially given that you are buying a problem for your cat, instead of treating a problem. A single application can be deadly for your cat. Since these products aren't very effective, some cat owners will apply it again soon after when they realize the fleas didn't care about the first application. Cats do not have the liver enzymes they need to break down organophosphates and so it builds up in their systems with each exposure. Eventually, sometimes after a single application, we will see these cats develop tremors, anorexia or vomiting, seizures, or imminent death. They require intensive treatment and care if there is any hope to help them. Unfortunately, the opportunity to recite this particular theme usually comes after the exposure, not before, and so the fleas and the over-the-counter products speak for themselves at that point.

The Domesticated Cat

Many argue that cats chose the indoor life and charmed us into it. They talked us into providing their meals, warm beds, and cleaning up after them. Perhaps those ancient felines in those early days of domestication miscalculated all the implications. They underestimated humans, not believing we could be as devious as they could be and probably never imagined once they had gained entry into our homes and enslaved us, we might eventually prevent them from ever leaving again. They traded freedom to roam, socialize, breed, sunshine, exercise, purpose, a natural diet for relative safety, readily available food, and comfort from the elements in a much smaller and less interesting world.

The health risks for indoor cats are real and there are many. We have already discussed some of them earlier in this book. To expand upon these, specifically regarding indoor cats, many feline health issues stem from two things: one is stress and the other is boredom. Now I know what you are thinking, my cat doesn't seem stressed. I am certain you are right; he is happy, has all the love he could want, the best foods, toys, and a bird feeder to watch the birds—what more could he want? He plays, he sleeps in the sun, he has a favourite chair and he gets along well with his housemates. And, he doesn't know the difference right? He has never been out, so how could he miss it? And you would be right about all of these things, except that the life of an indoor cat is not normal, because it is a fabricated life made by human beings in an attempt to keep their felines safe.

Yet, we have not quite perfected domesticating cats. Yes, thousands of years ago cats convinced us to let them in, but it was our idea more recently to keep them there. Cats are not stressed because of what we haven't provided; they are stressed because of what we cannot provide, the balance required for them to become their full selves, or to reach their full, natural cat potential. It is a physiological stress that occurs within the cat, even if we do not see obvious outward signs at first. This chronic physiological stress leads to behavioural problems in some individuals and causes elevated cortisol, which over the long term leads to inflammation and disease in predisposed individuals.

What does stress look like in a cat? The big four are house soiling, aggression, spraying or marking, and overeating. Other more subtle signs may include withdrawal from society (hiding frequently), reduced or excessive grooming (often producing lesions), frequent scratching or shaking head, failure to bury feces, reduced appetite, constipation, diarrhea, vomiting, piloerection (raised coat along the spine), eating things that are not food (pica), and catalepsy (spaced out, absent appearance). Remember this isn't stress related to not being loved well enough or because he is poorly treated, though this may occur in some homes. It is environmental stress related to indoor status and is seen most frequently in multi-cat homes. I consider this indoor status to be a sort of underground modern day stress like the bottom of a great iceberg. Many individuals cope well with it, but add to it other stressors like multiple cats and maybe some change in schedule or home, and individual cats may stop coping.

What are other stresses to a cat? They may include the following; overcrowding, dominance issues, a new home, a new family member or a new pet, absence of a family member or a pet, new odours or sounds, new furniture, rugs or house dressings or a new configuration of furnishings, home renovations, a new neighbourhood cat, boredom, and a change in routine among family members. These things construct the upper levels of the iceberg that now becomes more visible and may begin to cause symptoms related to stress.

I mentioned earlier that individuals weaned from nervous or nutritionally stressed mothers will be programmed to have a low threshold for stress primed already by cortisol when they were developing in the womb. They tend to have more inflammatory issues than other individuals and may display more frequent signs of stress noted above.

The boredom problem is a bit easier to understand. It makes sense. Some cats that are bored will overeat just like we humans, and this combined with a more sedentary lifestyle will lead to obesity, and then obesity will eventually lead to a whole host of other problems, more joint issues, and heart disease. Further, difficulty grooming can lead to painful mats, vaginal, urinary tract, and anal gland infections. Diabetes mellitus is also frequently seen. Other cats suffering from boredom will develop behavioural problems, for example, out of box elimination problems, over grooming, or destructive behaviours.

Destructive behaviours like scratching the furniture, rugs, and wallpaper are common complaints among cat owners. Cats lose their homes for this kind of thing. This of course is normal behaviour and so very difficult to treat other than to offer more attractive things in the right locations for them to scratch. Cats scratch in order to keep their nails conditioned. At the same time, it serves to scent their surroundings. The key to good scratching materials is to observe where they are scratching. If your cat is scratching your lovely couch in the living room that tells us a lot about what sort of substrate or material he likes best to scratch on; it tell us he likes his scratching post to be very sturdy, upright, and immoveable, and it even tells us his preferred location. Likewise, if he is scratching the carpet, we know that he prefers a rug or rope like substrate to work his nails and that he prefers a horizontal plane to scratch, and it tells us about his preferred location.

The next step is to try to train him off the objects he is using and train him onto a replica of some kind in the same or close to the same location. It is common for cats to want to do their scratching where he likes to spend most his time, and usually that will be close to where you like to spend most your time. To train him off the location you do not like you will need to use some deterrents. A citronella type scent may work applied to those sites. Double sided tape can also work. As well, putting an upside-down carpet runner in front of those locations can work. To train him to a new location like a sturdy scratching post, make sure it has the proper substrate that he likes and that it is substantial enough for him to consider it a better option. Rubbing it with catnip or spraying it with a catnip concentrate or pheromone spray will usually help.

Location, location, location is very important. Praise him when he uses it. Put treats upon it. The more he uses it, the more he will want to use it because it will smell like him and he will like that. There are many clever scratching posts and devices on the market today. My cats love the cardboard one with the ball that runs around the outside of it in a plastic trough. This is apparently the most fun of toys as well as a scratching device. One of my cats loves to scratch cardboard and books, so he has a piece of cardboard that I replace as needed taped to the floor. He also has a few big sturdy books (his favourite is a dictionary) placed where I know he will use them. He was using the dictionary anyway so I just gave it to him. Some of my clients donated (or sacrificed) furniture to this use with their own cats because they realized it was the best option in the end. It then becomes the cat's couch but he will still be very happy to share it with you. Somehow when we think of it this way, it becomes less of an annoyance and more of a solution that pleases everyone. Other cats prefer things brought in from outside, like part of a downed tree with thick furrowed bark. However, you may bring in some ants or other critters with these things, so be careful.

Keeping your cats' nails trimmed will certainly reduce damage due to scratching and may even reduce the amount of scratching he feels he needs to do. Nail trimming may be needed every few weeks, and more frequently in the kitten. It is best to start trimming nails as soon as you get your kitten, so that they will be accustomed to it and it will not be a difficult stressful thing for you to do. In fact, have your veterinarian show you how to trim your cat's nails on your very first visit. For those cats that don't like it, a few nails a day could be a routine they may accept.

My cats don't mind so much but they don't like to be hunted down for it, especially Merry. So, I have several pairs of small clippers about the house close by their resting places.

Sometimes it is decided, usually due to destruction of furniture, that a cat must be declawed. I usually start this conversation by stating that this is illegal in some countries and is considered maiming your cat, so you really need to consider all options before going ahead with surgery. If the cat could choose he would definitely say, "No, thank you." However, if it is done early before his nails are thick and almost bony at the base, then they heal quickly and the cat will not look back. It is actually quite a fun surgery to do if you can forget what it is you are doing. The nail is removed and with it the first little bone that the nail grows out of, so it is an amputation. Normally only the front paws are declawed and the back nails are left for defense in case he finds himself in need at some point. Outdoor cats should not be declawed. Most clinics will keep their declawed cats in the hospital for bed rest for two nights and then send him home with some pain relief for a few more nights. Remember when a cat is declawed there are five little wounds on each paw that need to heal and all the while your cat must also get himself around on those paws to eat and eliminate and find his bed. You will need to put all his necessities within closer reach for him so he does not have to go far while he is recovering. Usually, a cat will restrict his own activity while he recovering, but don't allow him to jump or climb as he may injure himself and set healing back a bit. It does take some time and he will have a shifting lameness for a period of time after his surgery. He will hold his paws up at times. He may milk it when he sees the attention it gets him as well. Declawed cats will still go through the motions because they will still want to use their paw pads to scent their environment. I occasionally see a cat with phantom pain that persists for months, sometimes longer, but when you palpate the paws there is no reaction. So, it is difficult to say if there is pain there or if he has just developed a gait that attracts attention. It is best to give him the benefit of the doubt. Ask your veterinarian. He may need some short term help for pain.

There are nail covers that can be applied by using a surgical adhesive when declawing doesn't feel right, but I have not found this to be a good long-term solution. They are cumbersome to apply, they fall off as the nail grows, and need constant replacement. These are great for when you are travelling with your cats to other people's homes where there may be other pets or children and there is some fear on the part of your hosts about scratches from your cats. This is what they are good for—short-term use.

Another behaviour that causes some stress for some cat owners is when cats jump up on the kitchen table or counters. Of course, this hardly seems sanitary, as we know where their paws have been. It is best not to encourage this. Motion detectors with either a citronella spray or horn attached will work well, but they are not perfect because they will go off when anyone approaches and so need to be turned off and on again. Tin foil or upside-down mousetraps can work as a deterrent. Upside-down carpet runners are another one that I like to use for many things to deter cats from a location. Long-term use of these things does get tiresome and are often abandoned when the cat appears to cease the behaviour. I'm not sure there is any permanent way to keep your cat off the counters and tables if he is inclined to be there. I think a cat learns very quickly not to do it when humans are around and that it is fair play when they are not. This is why I don't even recommend a water squirter. They are very effective to keep them off when you are home but they know that you drive it, and when you are not there, they are safe from being sprayed. That is why something remote from you works better, like the motion detector devices already mentioned or, dare I say it, a shock mat.

In my household, when I discovered the cats were sitting on my kitchen table for a good part of the day, I decided to just put up with it because I know the best windows for watching the birds are in the kitchen, and I know cats love to be up off the ground and elevated to watch their surroundings. It is normal. I use table clothes to keep my table clean. My cats also love place mats, so they sit on the place mats and I remove them when we eat instead of the other way around. This is my solution to a problem that they don't see as a problem. They don't necessarily care that it is the kitchen or the surfaces where we prepare and eat our food. It is the vantage point that interests them, at least initially. Some may scavenge if they happen to be rewarded by an open butter container while they are up there. But that is not usually why they start jumping on the counters initially. To eliminate scavenging, put away food that they may be interested in. Do not give them human food or treats at the table—that will just teach them to look for it there. If they are getting up to drink water from the sink then they may be telling you they like very fresh cool clean water. Fountains are available, and I find cats that like sinks, and they like fountain water even better because of the cool way it circulates. Keep your fountain or other water bowls clean and fresh every day so that it tastes good to the discerning feline. Again, usually it is the vantage point that appeals to them. Providing better options for them in some situations, like climbers for example, will often work. Moving the birdfeeders and placing them in another available window with a climber might work. The problem there is that I also like to watch the birds when I sit at the kitchen table! I am very motivated to keep my cats happy because I know what it looks like when there are environmental deficiencies.

Long-term physiological stress due to environmental deficiencies causes elevations in cortisol levels, which eventually contribute to or result in a number of effects. Simply put, it forces the body into overdrive, the immune system becomes stressed and the cat becomes more susceptible to infections and problems associated with inflammation. The same is true of humans living stress ridden lives. Of these, in cats, the most common include urinary tract issues, colitis, and various inflammatory processes of the pancreas, kidneys, and other organs. Triaditis is a fairly common disease entity in cats. It is concurrent inflammation of the pancreas, liver, and gastrointestinal tract and is always on the rule out list for cats with recurrent gastrointestinal symptoms such as a poor appetite, soft or runny stool, and vomiting. Again, it is inflammatory in nature.

Of course we see these disease entities in outdoor cats as well because stress and boredom are not only risk factors, but they do not occur in outdoor cats with the same frequency. Outdoor cats, for example, tend to have very good teeth in comparison, with little gingival inflammation even when they hit their teens. They do have more broken teeth that may need attention, but will have fewer resorptive lesions (painful dental decay that occurs at the gum line) and less painful gingivitis and periodontal disease than indoor cats. Part of this is due to the fact that there is more opportunity to hunt outdoors. However awful as it is to contemplate, chewing on the hides and tiny bones of mice will help keep the teeth clean, but stress plays a role here too. Often all we see is inflammation in an indoor cat—no tartar, just inflammation—and sometimes no other cause can be found. These individuals seem to have more plaque intolerance. This can be a quality-of-life issue, because many of our cats have painful teeth and gums from a very early age and we don't know about it unless we look, because they don't tell us.

Toothaches

One thing I am absolutely convinced of is that cats have a very high threshold for pain. A cat will not admit to tooth pain until he literally has abscesses growing in his tooth roots. Only then will he have difficulty eating. He may at that time drool or rub the side of his face, but not before. You may notice some bad breath, but you will likely think it just normal for a cat's breath to smell. You will not feel inclined to check how his teeth look because he will not like it, especially if there is pain there. And because the pain grows gradually along with the tooth decay you will not notice his gradual loss of energy and playfulness and you will think he is just finally maturing. Wrong!

Dental disease is by far the most common health problem we see in veterinary medicine. It starts early, usually around two years of age for indoor cats and later for outdoor cats, and it progresses over the years. It is always a shock to the cat owner to learn that their cat has tooth pain and until they actually see the affected teeth they cannot believe it. It is a quality-of-life issue as well as a health issue, because there is chronic pain. Imagine a canker sore, or multiple canker sores in your mouth that never go away but get a little worse with each passing week. That is what our cats feel when they have early dental disease. Most of us can relate to the idea of a canker sore, then multiply that by 100 if you wish to imagine the pain of an abscessing tooth, which is the kind of pain that penetrates deep into the bone and throbs and begins to permeate your entire existence. Bad teeth eventually lead to other health concerns, the most common of which is kidney changes related to low grade but chronic infections and eventually serious life-threatening kidney infections like pyelonephritis.

What we see when the affected teeth are treated, sometimes by extraction, is that the cat becomes himself again, and is more social and playful again without the chronic tooth pain. Looking in those painful mouths with swollen pitted gums and teeth entirely encased in thick brown tartar, I sometimes wondered if these cats were silently screaming all along and we just couldn't hear them with our insensitive human ears. But it is far better to prevent it, than have to treat it later. Certain dental diets can help with this and other dental preparations including a toothbrush and enzymatic toothpastes. There are lovely little toothbrushes angled nicely to reach those teeth in the back. I know one client that uses an electric toothbrush on his cat and others that use the edge of a facecloth instead of a brush. Plaque is the sticky material that develops daily on the teeth, each day adding a new coat of it. Within three days, the sticky plaque becomes hardened to tartar and with each three days this coat of hardened tartar builds up alongside the new plaque. Daily brushing works best to prevent it, and even if we are brushing every three days we are usually staying ahead of the plaque. Remember that as humans, we brush our teeth at least twice daily, floss, use mouthwash, and we still see the dentist twice yearly for professional cleaning. It is not reasonable to think that eating dry food and a few tartar treats alone will keep our cat's teeth clean. Twice yearly physical exams are important to monitor dental health properly and can help avoid dental surgery. Your cat is not going to tell you so let your vet tell for him.

If you start with the young kitten, brushing the teeth can become a routine that a cat can even enjoy. You start slow with the right supplies and equipment and you do it every day if you can, but twice weekly at least. We had a patient named Calvin, a big king of a Maine Coon, who spent a fair bit of time at our clinic as a boarder and he came to us with his overnight bag full of his toys, his brush, and other belongings as well as his toothbrush and toothpaste. We brushed his teeth daily and he loved it. He actually opened his mouth to have them brushed. It had become part of his daily routine from his first days with his family. He would jump up on the bathroom vanity while his owner brushed his own teeth and then he would have his brushed. Calvin was one of my all-time favourite cats. He always pretended to be annoyed with us the first few days at the clinic; he would snub us, but then he settled in and engaged us instead of us trying to engage him. Calvin was a handsome boy, and quite mischievous. Calvin was also a diabetic cat and his family took amazing care of him. They regularly checked his blood glucose at home with a glucometer and also checked his urine glucose with little test strips placed in the litter. They monitored his drinking and urination patterns, his appetite and activity level and charted everything every day. His family had his diabetes treatment very fine-tuned. They made themselves experts on this disease. It was impressive. Calvin lived a very long and full life as a result.

One of the secrets to a long healthy life is good teeth. Your own dentist will even tell you that. And you want the best for your cat I know. I remember an elderly client I had many years ago before I began my feline practice. She came to me with an older cat. She'd been seeing another veterinarian for years, but had recently moved into a senior's complex and was closer to me at that time. Unfortunately, she came because it was time to put her darling cat to sleep; he had been failing for some time and was in kidney failure. She stayed with him as we performed euthanasia with him lying in her arms as he went to sleep.

Afterward she wanted to look at his teeth because she was curious about their condition. She cried when we looked at his mouth; he had such advanced periodontal disease and huge cliffs of tartar protruding from the teeth. Clearly he had a lot of pain there for a lot of years. And she was angry because no one had ever told her through the years and years of regular veterinary care. She was angry and she felt very guilty. And I did not know what to say, because I knew she was right. Her cat's oral hygiene had been neglected year after year, and so his entire health had been neglected. She had noted the odour but had assumed her veterinarian was keeping watch and doing her best for her cat. I don't know why dental health has not been a greater focus in previous years, but make sure your cat's oral health is discussed at each visit.

Tightrope Walk: Finding Balance to Prevent Boredom and Stress

Providing balance to prevent boredom and stress in an indoor cat isn't easy but it can be done. It takes some effort. Some cats will require more stimulus than others. All indoor cats, however, require a measure of stimulus to avoid problems associated with boredom and stress. As we have discussed, health issues related to boredom and stress are easier to avoid and prevent than to treat. The idea of giving long-term medication alone and the theatrics, not to mention the expense that goes along with that should provide the incentive to provide a stable cat-sensitive environment.

Cats are very sensitive to changes in their environment. They love structure and a scheduled life. They may act out when changes are made quickly; for example, a change in the family's schedule or renovations going on in the home can be very disconcerting to a cat and he may let you know in ways that you many not appreciate.

Lots of toys can be very helpful. Switch them up to provide variety. My cats love the little fur mice best, but also enjoy anything with catnip in it. Hide treats so your cats can hunt for them. Laser lights and ping-pong balls will help cats get their exercise. Cat climbers and places to perch high with safe access are important and are especially nice where there are windows to gaze out of and bird feeders to watch, but they must be located where the cat is inclined to go. Cats can learn tricks if they are motivated by treats, and this can be a fun distraction for them. Some cats watch TV and might enjoy videos with fish or mice or other animals. Treat-dispensing toys that can be rolled around on the floor are a favourite as well. Large sturdy scratching posts that don't sway when they work them and positioned front and centre where your cat loves to be, are so important. Large logs can be used as well; cats love these. There are also cardboard scratching contraptions that cats favour. Big boxes that are left over from new appliances or other things and smaller boxes are great playthings for a cat; keep them and cut holes in them. I've had clients that built shelves in a stepwise fashion up a wall to a window and beyond where their cats can perch safely and watch from above. The occasional mouse infestation would be a blessing to a cat, truly. Your imagination is their only limitation.

Environment Enrichment: Important to Preventing Stress and Boredom Health Issues

Providing hiding places for your cats is an important component to enriching their environment, especially if they are timid cats or if you have multiple cats.

I think the overall best way to enlarge your cat's life is to allow safe access outside. This may be through a cat door to an outdoor enclosure, for example, or on a leash or loose in a secure fenced back yard. Underground electronic pet fencing is another option that involves a collar and underground wire. There are collars available that use G.P.S. signals to track your cat's movements and will alert you when your cat leaves the property. Many of my clients walked their harnessed cats on leashes every single day, either in their yard or on the street. They have to be acclimated to a harness and leash because they can feel constrained and vulnerable at first, so start with it indoors and let them gain confidence. And you need to be careful. You don't want to be approached by a large stray dog with your cat on a leash.

I met a veterinary student recently who was telling me about one of his four cats. After a day of exploring, they found him at the end of a bloody trail and by some means, possibly a car, he had broken a leg. After surgery and the fracture fully healed, he began to get curious about going out again but they were too nervous to let him out. They set up a large dog cage outside their back door amongst some shrubs and he waits anxiously by the door to go to his private outdoor room every day. They open the door, he steps out; he waits for the cage door to open and steps in. He loves it. He's safe, and he's happy.

This same student shared another story about his cats. Another of his four got out one day. They heard a great commotion and ran to the door. Opening it, the cat flew in with a bird in his mouth. The other three chased him around the house until he decided he wasn't going to be chased anymore. He deposited the bird on the floor and all four cats sat around it gazing as if in shock and amazement at this creature he had managed to catch. Three of these cats are strictly indoor cats and so this was quite an event for them. The little bird was unharmed and sat up a little dazed and they still just stared mesmerized. Perhaps they were wondering, What do we do now? One of the humans in the family scooped the bird up and returned him outdoors. The cats watched from the window. Perhaps we don't want them catching birds, but I imagine that was a stimulating and enriching day for the four of them.

I had the pleasure of bumping into one of my clients at an art show recently and she shared with me that her cat no longer is suffering from skin issues. We'd been treating her cat for anxiety-related alopecia (over-grooming and self- trauma to the skin and we felt it was stress related). She and her husband had spent a small fortune on a secure high fence in their back yard where the cat was safe and secure and no longer required any sort of anxiolytic to control her urge to pull her hair out. I was so happy they had made that investment in their cat's health.

There are many websites and resources available on environment enrichment strategies and I encourage all cat owners to check them out. It's important. My favourite is the Indoor Pet Initiative website (previously known as the indoor cat initiative). The only limitation to cats fulfilling their full potential is our imagination. It will be hard to recreate that thrill of just escaping for the eighth time, but we can improve the environment for our indoor cats and keep them safe at the same time.

The modern day feline symbolizes our own modern day habits and changing world. The modern indoor cat is a barometer warning humanity of the effects of our propensity to spend too much time indoors among other things. Most of our modern day cats lounge indoors: they eat too much, rarely exercise—it looks a lot like the modern human. And we modern humans are also developing inflammatory diseases and mental disease at alarming rates. Indoors at our desk jobs, surrounded by electronic essentials, we are developing charges in our bodies that are not normal. We develop positive charge indoors that free electrons in the soil and the great outdoors can neutralize. Add to this a lack of sunshine, fresh well-oxygenated air, and minimal body movement in the run of a day. It is little wonder so many of us have inflammatory bowel disorders, adult onset diabetes, heart disease, and so on.

Chapter 5

We cannot forget about stress, which is the current diagnosis for everything. But I'm convinced it is not just the stress we create living our lives. In addition, there is environmental stress that we don't even realize is working at a different level and underpinning everything. I'm not talking about mean bosses, or financial stresses, or meeting deadlines. I'm talking about where and how we spend our day. It is not normal to be indoors so much of the day, or all of the day, or most of the week. It is not normal to barely move our bodies or to be sedentary. Environmental stress is that underlying underground or underwater stress, like the base of an iceberg. It is often self-imposed in the case of the human.

We have asked our cats to accept diets that they are not built to digest. We have done the same thing to ourselves. We are eating diets higher in refined sugars, grains, and fats. Evolution in our bodily functions is not occurring rapidly enough to accept these changes for us. Our foods refined as they are, and our fruits and vegetables grown in monocultures depleting the soils and coming from far and wide, cannot nourish us the way they used to. We must add back our vitamins and minerals as supplements in order to come close to fresh garden produce of old. Modern day wheat for example, similarly challenges our metabolisms, its gluten content so different from ancient wheat that is almost extinct now. This new gluten in the hardy pest and drought-resistant wheat strains of today is like a poison to many of us, or maybe all of us.

With our changing diets and modern stresses we also (along with our cats) often need probiotics (and sometimes prebiotics to feed the probiotics) as well to help us maintain a healthy gastrointestinal tract. As with oral health, a healthy gastrointestinal tract is very important to our and our cat's overall health. Pesticides and fertilizers have damaged our soils and waters both at the point of their production and at the point of their use in the fields and orchards. In the future, we will understand more fully how they are affecting our bodies but it is already clear that they hurt us. It is all so confusing isn't it? Who knows anymore what is good for us and what isn't? The threat of GMOs, which are becoming common place, and the possibility of foreign produce being grown in human excrement are new concerns. With our own issues here at home with BSE (Mad Cow Disease) and other food borne health risks, one can become paranoid about eating anything other than what we grow ourselves in our back yards. I have known human individuals who lived the healthiest lifestyles possible, avoided fatty foods, ate organic as much as possible, and stayed fit, but still developed cancers and other health issues. The same is true for many pets I know. We can do all the right things for them and problems will may still develop. Life for a human or a cat is complex: our genetics, our environment, our level of fitness, our habits and lifestyle, our nutrition, our frame of mind, all count.

It is probably time to reinvent ourselves, or to tweak our thinking on certain modern lifestyle choices that promote sedentary habits and less wholesome nutrition. It is a result of ever increasing efficiencies, created by us, that have caused us to work more and more hours each day instead of fewer and fewer hours. Our devices constantly connect us to our work. We don't ever have to leave our work in this way and some of us don't. Of course we are mentally spent when we finally finish our dinner and plop into our favourite chair with a glass of wine to numb us. And even then we are often checking our devices for that connection. This is a perfect example of how progress is perhaps not a very good word anymore for some things. We can slow down that race to the couch where we tune out. We can relearn passion in preparing meals and staying fit and spending time together doing, rather than watching. We can turn it around again, and live a life worth living with intent—a rounder, more textured life.

For a cat, life has changed significantly in the last few decades. The rise of the indoor cat lifestyle, living on one side of the door, has changed our idea of the cat. The genetics of the cat however has not changed in that timeframe. In addition to this, the rise of the multi-cat household has brought other new challenges to the life of a cat. You could argue and I have done so above that humans have done much the same thing to themselves. We have largely become indoor sedentary humans and our genetics have not evolved to support this change either. Medications and all sorts of fad diets and whatnot of course exist to help us cope with health issues that arise due to our deficient environments and sluggish metabolisms, but in the end it makes more sense to allow nature to correct things for us when possible, with sun on our faces, feet in the dirt, and moving our bodies. This often won't be enough, but is a good place to start.

Chapter 5

The indoor cat lifestyle can work very well for some individuals, especially the single cat in a loving household where all his needs are met. He will certainly lead a sheltered life from the risks of the great outdoors and should have few to no health concerns if his nutritional needs are met and he has regular check ups. As numbers of cats increases in the indoor setting, it becomes more difficult to provide the space and other elements they need to be whole. The result is often a sick cat. The cat's body will tell us, through the language of disease, just like our own bodies tell us when our lifestyle is not quite right. His symptoms may manifest into behaviour problems, like voiding out of the box, which are especially intolerable to their human companions. The indoor life therefore is sometimes risky for the cat, and often leads to death from euthanasia.

Learning the Secret Language of Cats

Chapter 6:
Curiosity Killed the Cat: The Outdoor Cat

Cats are curious and sometimes this curiosity gets them in trouble, or even killed, especially when they venture outdoors. According to various sources, it's likely that the cliché, "Curiosity killed the cat," originated from English playwright Ben Jonson's 1598 play, *Every Man in His Humour*.

Outdoor Cats: The Real Risks of Allowing Your Cat to Go Outside

We cat lovers all understand the risks of letting our cats go outdoors. I suspect I have lost three cats to coyotes, so that is why my current three cats stay indoors all the time. We discovered recently through a friend's webcam that they are not coyotes at all, but coywolves, which is a crossbreed or hybrid. They most likely originate in Algonquin Park and migrated throughout the province into or close to residential areas. I still miss those three cats I lost to the coywolves. We still talk about George, Tom, and Gus; they were all beauties, close friends, and companions.

Tom was probably my all-time favourite cat. He was big, glossy, black, and ridiculous in his affectionate attentions to me. I can still see him jumping up on the bathroom counter for me to turn the water on for him. After passing his paws through it several times, he would put his head right in the stream of water, let it run down over his head and lap it up as it ran off his face. He was rescued from the very top of a telephone pole, sitting there like an owl, when he was a year or two old. I was eight months pregnant at the time, and I heard a cat crying and meowing that sorrowful extended meow of distress. However, it was too dark to see where the cat was and his voice seemed to bounce about like a ventriloquist. In the morning, I saw a black cat at the top of the telephone pole a few houses away.

First of all, I could not believe he had managed to get up there. Secondly, I was amazed that he was able to stay balanced there all night. A neighbour helped get him down; this was a very dangerous process involving a ladder and his shed that I could not bear to watch. Tom became mine when no owner could be found. He was the very same cat I'd seen on the roof of our porch a few nights previous from my daughter's bedroom window, but he was gone by the time I went outside to see him.

During Tom's life with us, he disappeared a number of times after we moved into the countryside; once he disappeared for over three months. I can still hear him approaching from the cornfield next door; his distinctive meow reported his long-awaited return. He flopped down on the porch beside me, in good flesh with a lovely shiny coat, and began to groom himself as though he'd only been gone a few minutes. I suspect he had found a second home; perhaps he'd made a timely escape to come home to us. I still hope his final disappearance resulted in his return to his possible second home and his second family just decided to never let him out again.

We came across George as a very tiny kitten abandoned deep in the woods. My kids found him when we were hiking. He was alone and crying in a thicket of brush so he became ours. He sat in his bowl with the food as he ate for days; he was tiny and so hungry. He and Tom became fast friends. Tom taught him how to use the cat door. We were intrigued as we watched Tom poke his head through several times trying to get George to follow him out. Tom then went out and back in the cat door several times. Finally, tiny George realized it was safe and followed after him down into the basement. He and Tom walked side by side in tandem, keeping time in matched steps even around the corners, shoulder-to-shoulder and hip-to-hip. They slept in a swirl of cat, black and grey tabby, and they groomed each other so well until their ears were wet. We lost George after Tom. I think he went out in search of him.

Gus came from a client who was feeding about 16 feral cats in her back yard. Gus was born into one of the many litters the wild cats produced before they were all caught and neutered. The kids named him after Cinderella's little mouse friend. It took him years to trust the adults in the house. He'd come out for sour cream as a kitten and he would let the kids hold him and cuddle him and eventually he would purr. We rarely

Chapter 6

saw him otherwise until he was five or six years old. By this time, Tom and George had already gone missing, and then Gus ruled the house. He never did learn to trust my husband, but if the kids pre-purred him in those early days, then he would sit happily in my arms or in my lap. Later, Gus would nip my ankles when he was hungry. I loved him. He was my assistant and blood donor for the little flea ridden and anemic kitten I mentioned earlier. Gus was my hero; he was a lifesaver.

Gus went missing when I was away travelling. My husband and the kids heard wild crying one night and the sickening sounds of a life expiring. They called and called him but he didn't come and the next day he was still missing. I found his intact remains a week later in our neighbour's field next door. I am still not sure what happened to him. Perhaps the children had interrupted the fatal assault by calling repeatedly for him. Or perhaps the coyotes were after something else that night and Gus had found some poison.

The coyotes in our neighborhood have moved on for the time being, but it is difficult to accept this kind of risk again. I've heard so many sad stories regarding lost cats. It is a wonder anyone lets their cats out at all. At the clinic, we received so many faxes sent out with pictures and details of a beloved cats' last whereabouts; so many posters posted in the reception area. And it is a wonder that I would even suggest it. But we need to balance risk.

Going missing is not the only risk associated with the outdoor life. As discussed previously, cats that go outdoors will occasionally fight with another cat. This will often result in abscesses. A cat's bite can penetrate very deeply into tissue below the skin, which then injects bacteria from the oral cavity into the puncture wound. The bacteria love this location and will proliferate quickly there and create a larger and larger pocket of pus. The area will be painful and the cat may develop a fever and be lethargic. He may decline to eat. Sometimes the abscess will break open through the skin and a putrid smelling discharge will seep from it. Abscesses require treatment with antibiotics and thorough flushing of the pocket to remove the pus and eliminate as much of the bacteria as possible. Antibiotics alone at this stage will not be effective. The antibiotic cannot penetrate a pocket full of pus because there is no blood circulation within it that can carry the medication into its centre. In essence, the abscess needs to be lanced and flushed copiously. This is usually done under sedation. They respond quickly to treatment. Once an abscess is formed it is not possible for a cat to recover from that by himself. He will need veterinary care. Early treatment, before a pocket of pus has formed, will eliminate the need for sedation and will prevent that stage where the cat feels very unwell.

Giving medication to a cat is sometimes the only way to get them well again. Finding a way to get it into them can be a challenge. When the difference between a dead cat and a perfectly healthy cat is 10 days of antibiotics for example, this is one time you cannot let your cat choose. He has to take it one way or the other or he will die, or if not die, then he will develop secondary problems that may also require medication for a longer period of time. The sicker the cat is, the easier it is to give medication, so sometimes those first few days are not too bad. Then, the cat becomes ready for that next dose, because he knows it is coming, and then trouble begins. You know you have to get it into him and he knows he cannot let you on point of pride. So then the drama and theatrics begins.

The good news is that like everything else, there have been great strides made in pharmaceuticals for cats. Many preparations have been improved to taste much better and to come in smaller and smaller pills or liquid volumes. There is even a long acting cephalosporin antibiotic injection favoured for skin infections, for example, that may in certain circumstances be appropriate. Remember though that a veterinarian needs to be thoughtful about antibiotic choices reserving certain ones for use only when absolutely necessary. Antibiotic resistances are becoming a very serious problem for pets and humans alike. We have to be so careful to use them appropriately, at the right dose and the right amount of time and only when indicated. More and more medications have been formulated to be absorbed through the skin, others through the oral mucosa and do not need to be swallowed at all. A variety of special therapeutic veterinary diets now exist to treat such ailments as arthritis, obesity, chronic constipation, urinary tract disorders, diabetes, hyperthyroidism, and even aggression and anxiety. When the only option for what ails your cat is a tablet or liquid, fear not—it can still be done. I would prefer to give a pill over a liquid any day of the week. You pop it down and double check it

Chapter 6

is swallowed and it is done. With a liquid there is always a chance he will spit a good portion of it out, unless it tastes very good.

What I like to do when pilling a cat is to approach him when he's relaxed. I sit down close beside him, then I tip his head up gently but firmly by scruffing him high on the neck (with a smile and a greeting) with one hand. I talk to him for a second or two while, with one finger of the second hand, I gently push his bottom jaw down and instantly pop the pill as far back as I can with another. Sometimes it is a two-person job to keep the cat in place, but if you approach it like it is not a big deal, then the cat may not care very much. Try to be casual. What is even easier is if you have a pill gun (keep it behind you until the last second). The idea of a pill gun sounds awful but they are very efficient at getting the job done quickly, and they do not hurt the cat if used properly (your vet can show you how); and therefore, they reduce stress associated with giving medications. I love a good pill gun. However, I know from time to time that there will be that cat that just will not see it your way.

One of my most favorite clients had a cat named Pita, also known as, "Pain in the ass". Pita was a tiny but most demanding and challenging Siamese cat who never shut up, according to my dear client. I cannot remember what was wrong with Pita that first time I met her but I prescribed medication for her and instructed my client on how to give it. She just glared at me as though I was crazy and then challenged me to try it myself. I was not the least bit concerned about doing that and had intended to give the first dose in any case but as I went to give the tablet I advised the owner as I always do—that in the clinic setting it is often easier for us to give medications than at home because they are a bit nervous and more timid in the clinic setting. Also there is no place to hide, which gives us a huge advantage. Pita was true to herself and her name and I went through several tablets, discarding those that had been moistened and spit out before I succeeded on about my twelfth attempt using a pill gun, a tightly wrapped towel around her, and my technician holding her. A cat had showed me up in my own practice. It wasn't pretty. Sweating and bleeding, I apologized for having to send her home with any medication at all, because I now understood what she was up against. She asked if we could wait until she was a little bit sicker and it was actually tempting. She did find a way to give the medication without any stress at all by wrapping up in a little bit of cheese slice. So, there is always a way, we just need to explore sometimes.

Another client was very excited to call me and tell me that she had discovered an ingenious method to give her reluctant cat her medication for hyperthyroidism, but she did not feel it was working yet. Then a week later, she called to tell me she had found a dozen tablets deposited behind a pillow on the couch. Today, there are moldable treats available for pilling purposes. You can press the tablet right into it and mold the treat around it and most cats adore these treats and will be quite happy to take their medication that way. You give them a couple, one with the pill in it, and they beg for more and are happy to take the next one when it is time. In fact, they will tell you when it is time. Cats with diabetes also will sometimes let you know it is time for their insulin injections. Clients are always nervous about starting insulin injections but cats really don't seem to mind, and I think because it makes them feel so much better that they quickly learn that it is the needle that makes them feel good. Modern veterinary diets designed for the diabetic cat help in reversing the disease for a lot of cats. Treating hyperthyroidism in cats today does not even require medications in some cases. There is a complete veterinary diet very low in Iodine that will bring thyroid activity back into the normal range. Now that is progress, but for the cat with the large abscess you will still likely need oral medications and will need pill treats to complete the course as directed, especially if you have difficulty giving medications. And you will also need to clean the wound daily even after your veterinarian has lanced and flushed the abscess. This is so that drainage can continue to occur and that healing occurs from the inside out. Your veterinarian will give you instructions for care and schedule recheck appointments to make sure the abscess is resolving.

In addition to the formation of abscesses, bite wounds can transmit viruses like Feline Leukemia Virus and Feline Immunodeficiency Virus. Remember when your cat goes outside, he may share space with cats in the feral communities where these viruses may be rampant. Poisonings, accidents, traps, predation, catnapping, foul play, fleas, parasite infestations, and the most terrible hit by a car (we call it H.B.C. because it is such a common occurrence) are all risks of the great outdoors.

How can we balance the risks of indoor versus outdoor life for our cats? We need to assess the pros and cons to decide this. As a veterinary health care provider, I will always prefer to see something prevented than cured, because sometimes there is no cure. This ideal seems to put us at a crossroads of sorts. At this juncture, I believe that we need to let the cat choose and then accept the consequences of that for his sake. For many, this may seem like poor parenting, much like sending toddlers out to run in the streets, but it comes from years of observation and assessing risk. And they are not toddlers—they are cats.

If a cat is disinclined to go out, then we keep him in and remember to provide as much environmental enrichment as we can including the important space he needs. This may mean limiting the number of cats you can have in your home. How much space is enough? It makes sense to me that if a cat in a feral outdoor community requires a certain amount of space in his inner domain where no unauthorized cat may venture without consequences, then he will require that same or close to the same amount of space indoors. Home ranges can overlap and there will be communal areas within the overlapped regions but a dominant cat will have an inner domain within his marked territory that is his alone. He may allow key individuals access into this space.

On the other hand, when cats go outdoors, a cat can choose which individuals to allow in his or her territory. It is interesting to note just how large a home range is for a cat that goes outside. Think of the home range like a dart target board with concentric circles and a bull's eye in the middle. The circle just around the eye is the inner domain; it is his private personal space. The territory is outside and the home range is the outer ring. He will defend his entire territory from roving cats or cats of other colonies. Part of his daily routine will be to walk the perimeter surveying and marking where appropriate.

The home range does differ between the feral cat and the pet cat; the feral's range is much larger. One study out of the University of Illinois (Jeff A. Horn, Nohra Mateus-Pinilla, Richard E. Warner, Edward J. Heske; The Journal of Wildlife Management No. 5 Vol. 75, July 2011) suggested a home range of over 1,300 acres for one male feral cat in the study. But even a male pet cat will have a mean range of 4.9 acres and some will go significantly further. Females tend to stay closer to home, usually staying within half of an acre of home. Within these home ranges there will be areas designated for elimination, sunbathing, hunting, sleeping, and resting. The Illinois study also showed that feral cats are more active, are more nocturnal, and changed their habits with the seasons. It also showed that even owned cats can have serious impact on the environment in terms of competing for prey species. It highlighted risk of spread of disease from feral communities to family cats. All further evidence that population control and neutering and spaying our cats is so important.

The inner domain or personal space within the home range is much smaller—exactly how large is not known. I am unaware of any study that has been done to determine this but my multi-cat indoor patients suggested in many cases it is larger than the average home. His territory extends beyond the home and often the yard as evidenced by some cats' wild reactions to cats they see through the window visiting the perimeter of their homes. His reaction however may show up only as an unacceptable behaviour. A cat's home in this case must be his home range, his territory, and his inner sanctum or domain. As our pet cats are generally neutered and spayed, have litter boxes, and have all their food provided for them, a smaller home range suffices. We know from telemetry studies (following pet cats carrying tracking devices in their daily travels outdoors just like a G.P.S.) that home ranges like the ones stated above are smaller for pet cats than feral individuals. But is their inner domain or personal space smaller? If so, by how much? We do not know exactly what that number is. In other words, I do not think you need a home of 4,000 square feet to house two indoor cats comfortably, but we do need to consider space when we decide the number of cats in a household. We can increase useable space by providing vertical options for perching and resting like climbers, shelves, and window seats, and by including hiding spots like boxes.

If he wants to go out, then we need to understand and accept the risks of that too. He likely won't only stay in the backyard, because these telemetry studies have shown us they may actually go quite far. It is usually not such a challenge to keep a kitten in as they can find mischief anywhere at first. But once going out becomes an objective, there is no stopping them. I believe the longer we can keep them in, at least until their first birthday, the smaller their home range will be, but they may also be less street smart. If you live on a busy street, then I do not think it is advisable to let him out at all. You will want to consider other options like outdoor enclosures, which I love, and also harnesses and walking your cat preferably in a safe fenced area. However this will often just give them a taste of the wilds and create a situation where he will try to escape when he can. Alternatively, invisible fencing options do exist for cats as they do for dogs. It does require wearing a collar but these have been fashioned for cats and can work very well to keep them close by.

If your cat is going out then I would gently suggest that feral cats in your neighbourhood could benefit from your help to keep their numbers down, which in turn reduces disease and parasites within your neighbourhood, which your cat may be exposed to. These cats need our help. I think of the unspayed female feral cat having litter after litter, barely able to keep flesh on her own bones. The female and her offspring form the nucleus of many feral communities, which are often matriarchal, staying close to where they were born. Eventually most of the males are sent off or wander off to find their own spaces. Catch and release programs exist and need our support. This will be an important part of keeping your outdoor cat safe from disease. Small contributions become massive and meaningful when all cat owners contribute in the name of their own cats' safety. There are many worthy associations in nearly every community that assist feral and barn cats with T.N.R. (trap/neuter/release) programs.

Sometimes there is no choice in it. The cat chooses. He finds his way out with the children when they come in and out. And to be very honest, I think this is a very beautiful thing. I love to see a cat lying on his front porch grooming himself or his head popping up through the long grass in a field as nature intended. Sometimes we have to make our peace with it. Wouldn't it be awesome if all cats could be equipped with a tiny G.P.S. (like the collars that are available) like microchips under the skin that could actually monitor their movements and position at all times? Then we could know where to go to call them for supper or for bedtime. They might even come the odd time.

Oh, the trouble that can be found out of doors. There is a classic story about a fearsome cat that once lived on our street that is legend. No doubt it has been told again and again and has morphed somewhat and no doubt it has probably even been written about in a book or two, so I will try to get it right. This cat, a very confident and good natured cat found himself at the top of a magnificent fir tree one day. How he got there isn't clear. Maybe he followed a bird or squirrel, maybe a dog chased him up it (the owners won't admit to this), or maybe it was the sort of tree that had perfectly spaced branches that made climbing it impossible to resist for such a curious cat. In any case, he got very close to the top and would not come down. For several days and nights, the neighbourhood could hear his bellow from the top of that tree. Eventually, it was decided when he could not be coaxed down that the tree should be bent to such a degree that the cat could make an easier leap from the boughs to the ground. So, a rope was tied around the tree close to the top. I picture a huge ladder or a bucket truck being needed but I think what I heard was a lasso, a huge ladder, and a truck was used to tie this rope around the top of the tree. Then the rope was stretched, which pulled the tree down with the use of a tractor. When the cat still did not leave its boughs; it was decided that the rope should be cut. And so it was cut. On the first thrust back of the enormous tree trunk, the cat clung tight and cried out his shock and surprise at their solution, suggesting there must be a better one. But on the second swing, this fine cat was catapulted, not inelegantly (but for a screech), out of that tree into the back yard 100 yards away and landed on his feet. From his landing spot he sauntered off nonchalantly, seemingly unscathed, without a glance at his rescuers and went back to the barn to find his food bowl and have a nap.

My husband often reminds me of a cat I saw early in my career when there was no such thing as an emergency clinic. Sometimes when I would come home and start telling him about an interesting case, I would see that grimace form and he would simply say, "Do you remember that cat…" He didn't even need to finish the story about this poor cat that he had helped me with so many years ago. I knew the cat he was talking about amongst the many that he has helped me with over the years. This cat was found one Saturday evening collapsed on his doorstep after having been missing for a week or so and I was paged in to meet the client at the clinic. My husband often assisted me on those weekend evenings when I was on call. This cat had obviously dragged himself home. His lower jaw was badly broken and askew and he had gravel and dirt ground into his road-burned face and body. There was a terrible odour coming from within and around his mouth. His pelvis was also broken and he had deep wounds around his tail and rectum that had attracted flies to lay their eggs there. Maggots swarmed the area spinning in and out of little tunnels they had formed in the tissues. This was my initiation into maggots. We had seen many shocking things as students in veterinary school; in fact, each day brought some new kind of shock until I thought all the shock was gone. But somehow I had missed maggots. My husband had not had his sense of shock numbed in any way and so he gained a new respect for the work I did that evening. With his eyes half closed and swallowing down his nausea, he also gained a new appreciation for cats in general. Maybe this cat had lost two lives that week. He was weak and dehydrated, but when offered food, he ate and ate even though his jaw moved in different directions. He had such determination and tenacity to live. He'd made a miscalculation, thought he could make it unscathed.

No doubt, he had been hit by a car and had been lying dazed in a ditch somewhere for a day or two before he managed to get himself up and make his slow way home. We hooked him up on IV fluids, tended his wounds, gave him pain medications, and began some antibiotics to treat his infected open wounds. We took x-rays and determined that his pelvis, though broken, would heal on its own and thankfully no other fractures or problems were found. I was able to remove most of the maggots using local anesthetic and the next day when he was rehydrated and stable, I gave him an anesthetic to wire his jaw back together so that it could heal, cleanse his wounds again and remove the rest of the maggots that had burrowed deep into his flesh. He looked so grateful and happy, and when he woke up, he was already feeling so much more comfortable. This cat healed very well and looked like a million bucks a few short weeks later. You see, it really is hard to kill a cat; they are amazing healers.

When our black cat, Tom, went missing many years ago for the final time, we put posters up everywhere with his picture and description and offered a reward for his return. It was interesting when we received such an enthusiastic response from these posters. Everyone seemed to want to help, which was lovely. However, no one could be convinced that the cat that we had lost was a shorthaired black male neutered cat as posted on the poster. On three separate occasions, we were called to a home who had claimed to have found a cat fitting our description to find an orange tabby cat in one instance, a brown tabby in a second, and a grey long haired cat in the third instance. They were all lovely cats, all needed a good home, and all were lost or stray, but none of them were our Thomas. I think perhaps each of these kind souls felt it unlikely that we would find our Thomas and hoped we would take the stray they had found to replace him. We hadn't been ready for that and the idea especially upset the children.

There are some things we can do to help make a safe return home more likely. Tattooing cats has become out of fashion in favour for microchipping. Tattoos are problematic as they are messy, often illegible, and some feel that they disfigure the cat, but if nothing else, they clearly say that this cat is loved and is part of a family. I wonder if perhaps if we designed some kind of small beautiful tattoo for the ear of a cat that was universally acknowledged to be a sign to look for a microchip that it would be helpful. I wonder if a tattoo like that would ensure that more people might be inclined to have these stray cats checked for microchips and then they could be returned to their families. Tattoos are in vogue after all.

Wearing a collar with identification would be ideal if cats didn't ditch them so often. They need to have a break-away design so they don't strangle a cat when he gets himself into a tight spot. I like the idea of the underground pet fence for cats. This is the ideal solution. In fact, I think if I did go back into feline practice, then I would have in my pocket a prescription pad inked with the word, "Fence," because that would work to keep both the indoor and the outdoor cat safer. A high above-ground fence is an option but may be costly. The underground fencing is a newer idea for the cat but at least one brand that I am aware of has been adapted for them. After training, many cats will need to continue wearing their collars in order to keep them within their boundaries. Others learn and accept forever where they are allowed to go and where they are not. Cats learn the boundaries quickly. Like the kitchen table problem, I suspect that as soon as the collar is off without the reminder beep (before the little correction zap) as they come close to their permitted perimeter, then they would assume someone else's backyard is theirs for the taking. But, wearing a collar when you go outside seems a small price to pay for enlarging your world. These newer collars can also be programmed to open and close pet doors so they can come and go as they please. Remember an underground pet fence will work well to keep your cat confined but will do nothing to keep others off your property. You need to make sure your cat has readily accessible and safe hiding places where he or she may flee if there is danger. There is also some risk that he may lose the collar or that he may snag it on something and become caught but I think overall these risks are very small. If he loses it, it will be within his fencing, not outside of it, and it is unlikely that on a single collarless day that he would venture beyond his range set by you. These fences can be used indoors as well with the use of little monitors plugged into the wall, which could be hugely helpful to keep a cat off the counters or away from the Christmas tree, for example. I use it for my dogs that enjoy the kids' lunches before they are put in their lunch sacks and always think fresh pies are made for them.

Cats do return sometimes; our cat Thomas returned on multiple occasions, as did Molly—both after extended periods when we thought they might be missing. With Molly, I had been particularly worried because we were about to move. She'd been gone over a month and we were moving in a few short days. She came home and moved with us a few days later. She was emaciated and stressed and I suspected that she may have been locked in someone's summer cottage or garage and perhaps she had been surviving on mice. Cats can survive for several weeks with minimal food, but they do need a constant source of water. Cottage country was just a few streets away. Cats have very good homing instincts and can usually find their way home if they are unobstructed by circumstances or injury. When they go missing, it is often because there is a new home involved and they head off looking for their old home and then may become lost.

Some cats find new families and stay for just a period of time or even forever. I remember one adult cat that accompanied a young lady to my clinic for a checkup and vaccines. The girl happened to mention that the cat had just shown up at her home a few weeks prior and had moved in. They'd made an immediate bond and she was head-over-heels in love with this cat. I perhaps unkindly suggested that we check for a microchip, because we give them so cats can find their way home, and she agreed that we should do that. We found a microchip and the owner, but the new owner was devastated. We called the owners from the clinic to let them know we'd found their cat. They lived just two streets over from where the cat was found. I don't know if the cat's original family was just insulted that the cat had chosen a new home almost in their back yard, or if they didn't care so much for the cat, but they said quite simply that the new owner could keep her and hung up on me. So, it was a happy ending to what could have been a very sad story for this girl at least, and for the cat as well if she'd had to go back to her original family who clearly didn't love her as much. Perhaps this cat didn't like her two housemates and decided she wanted a fresh start. This is what I mean by a cat prospering in their natural setting. They can make things happen.

We had another client desperately looking for her cat. He'd been missing for a few days already and she feared the worst. We put her poster in the waiting room and faxed it to all the surrounding clinics and Humane Society but heard nothing back. A few days later we got a call from a clinic 400 kilometres away and they said they had one of our client's cats. They had traced him back to us through his microchip. We never figured out how he managed to get so far away in just a few days, but we decided he must have hitched a ride on a truck or someone's car, or had maybe fallen asleep on the back seat unbeknownst to the driver. Maybe someone took him and released him. This is when you understand fully how little we understand about the language of a cat. But it is also the mystery of a cat that attracts us. In any case, he found his way home. His very relieved family was happy to make the trip up north to collect him, and was also happy that they had decided to microchip him. Their only comment on getting him home again was that he smelled odd, like pine cones, for a day or two and that he slept for two days straight. Microchips do work.

Cats who have spent time fending for themselves outdoors often develop the talent for charming humans. I knew a cat by the name of a perfume that had picked her owner at the back of a skating rink, just as fall turned to winter. Another cat by the name of a racecar picked her family by watching races at the local stockcar racetrack. Daphne Du Maurier followed a client home from the library, Scarlet from the cinema, and Spinach from the grocery store. Some just show up for dinner and assume they are expected and long anticipated. Even cats adopted from the Humane Society engage humans they are interested in, after having ignored a dozen other visitors. We think we choose them, but the cat always chooses. Perhaps they are able to look deep into our souls and know we need them and only them or perhaps they see what they need in us. Perhaps it is simply a sympathy of souls that attracts them.

A Note on Dominance and Territorialism

It is true that large colonies of cats will form loose associations in feral groups. It is also true that very little argument will erupt in these groups, unless a new cat moves in and order must be renewed or the new cat flees. Yet, these cats know where they are in the pecking order. They know who is boss. And they know what a flick of a tail or a slight arch of the back means. Dominance is subtle in cats. They choose to live peacefully together and they are reasonable. They don't need to fight to maintain order.

Felines choose between these larger numbers of cats that they wish to stay close to, just as we humans do. They can easily avoid a cat that bothers them, or threatens them repeatedly, by just keeping a safe distance. They understand how territories overlap and where the inner core is where they are not permitted, and where they are welcome to spend time together and where they are not. There are no walls or windows to constrain their demonstrations or to prevent their finding their peace when they need to.

In the multi-cat indoor scenario, cats lose all control of their environment. If they don't like their housemate or the cat they see outdoors is provoking them, they cannot do anything about it. I can just imagine their frustration. It is always very sad when a pet dies. However, it is a common realization after the fact, that the remaining cat is overjoyed to be out from under the paw of his dominant housemate. Cats that owners have always assumed to be aloof, antisocial, or skittish suddenly are strutting through the house with confidence and not a care in the world, purring, and attending to their owner the way they always wanted to.

I'm sure many of you have also seen a housecat run from window to window screeching an almost blood-curdling scream at a cat prowling outdoors. Our Watson does this when a neighbour cat comes into his line of vision. It does not mean that he would fight with this cat if he met him or saw him out of doors, but he would make his feelings known about trespassing on his territory. If Watson had been able to deposit scent where he felt appropriate, and was able to make his patrols of the periphery, the neighbour cat probably wouldn't have dared. Eventually, the cats may even become friends.

So subtle is a cat's dominance that we often miss it altogether. When I ask a client whom they feel is the most dominant of their cats, they are often puzzled by the idea and feel that all their cats get along very well, groom each other, eat together, and sleep together. But there is always a dominant cat. He may share resources very amiably, like his favourite chair for instance, but he will have it when the sun is shining upon it, or when you are sitting with your afternoon tea.

Sometimes you can get an idea of who is dominant by observing who is out and about most or who uses the entire house—he is the dominant cat. The cat who uses little of it or who is off on his own most of the time is likely the subordinate cat. That does not mean that they do not share the same space at times. Even cats that pile on top of each other at naptime have an established hierarchy that is understood between them. Dominance issues can change however when a cat reaches social maturity; it is all up for election and there may be some discord until things are settled again. There is such a thing as a bully cat though—usually he is created by a subordinate's reaction to him. If the subordinate is very frightened of this cat due to previous unhappy encounters, then his extreme response of fearfulness (like victim mentality) will stimulate further aggression on the part of the dominant cat. This is a cat that enjoys his dominance and exerts it willfully.

I think it is very helpful to gain some insight into the social hierarchy among your cats because it can help put certain behaviours into perspective. It may help us to know what strategies to put into place for the cats on the lower rungs; for example, where to put an extra litter box, or where to put a box as a nice hiding place or quick escape in case of an incident. If one of your cats is hiding a great deal, then he is likely a victim of dominance or territorial aggression and needs some help. He will feel pinned down even by subtle cues from his housemate or mates.

Chapter 6

He will fall in line. At the same time, if you create the right environment for him, providing more places he can go to for safety and leisure could enrich his life. Putting a bell on the dominant cat is very helpful when there is discord or inter-cat aggression because this device will alert the less dominant cats to his approach so they can make a quick escape or find a safe hiding place and take a nap.

If a cat chooses to stay in, then remember it will be more work to keep his or her environment healthy, and a big part of that is ensuring that it is not crowded. It would be very lovely even for the timid cat to at least have the benefit of an outdoor enclosure of some description so he can get his feet dirty and investigate the great outdoors from a safe vantage point. I believe this would be enriching for even the scaredy cats that at first might not like the idea.

Vaccine recommendations differ for the indoor and the outdoor cat, as will deworming and flea prevention protocols. You will want to discuss this with your veterinarian before making any decisions and be sure to follow his or her guidelines closely to ensure your cat is as safe as possible.

Yes, there are lots of serious and sometimes deadly dangers outdoors for our curious cats. Fighting, abscesses, dread viruses, cars, predation, broken limbs, catnapping, poisons, fleas and parasites, and M.I.A.s. A cat will spend his nine lives outdoors, but it is decidedly a richer, fuller, more interesting life for our cats as well. It is a life better suited to their natures. If you do let him outdoors, you will be nervous every time you let him/her out, but you will also smile when you see him romping in the grass and stretched out or curled up in a sunbeam on the porch and knowing he is unfettered and free to be who he is. There are some measures that we can take to keep them as safe as possible including keeping current on vaccines, deworming, checking him over regularly for wounds of any kind, checking for ticks, and by making sure he has some form of identification so that he can be returned to you if he does go missing. Better yet, consider underground pet fencing, above ground fencing, or an outdoor enclosure. All in all, when you watch your cat outside, it will remind you that you too can enjoy the great outdoors and feel the same freedom and enjoyment as your cats.

Louis
Photo by: Madeleine Teed

Chapter 7:
There's More Than One Way to Skin a Cat

Basically, the meaning of the cliché, "There's more than one way to skin a cat," is that there is more than one way of achieving an aim. This chapter discusses the problem of cats eliminating outside of their litter boxes and how this can be detrimental to the cat, sometimes even leading to the euthanasia. This chapter will create awareness of this problem, discuss solutions, and show that there is more than one way to prevent this problem.

Elimination Outside the Box

Inflammatory problems of the urinary tract and colon, or chronic interstitial cystitis and colitis are of particular concern because they often will result in euthanasia of a beloved pet. Because of this likelihood of euthanasia, I consider these two health issues deadly—potentially far more deadly than all the risks of a cat going outside. The most common presentation for these cats in the veterinary clinic is urinating or defecating repeatedly outside of the box. When there is inflammation, there is pain, so a cat will quickly associate pain with the litter box, and he will be reluctant to go there to eliminate and choose other sites instead. Often after the cause of the pain is gone, the cat will continue to void out the box since he has learned he even prefers it. This isn't something that most humans can tolerate for long, and once elimination is the problem, it becomes difficult to find a cat a new home.

Eliminating outside of the box is certainly very difficult to deal with and when stress is a factor, this problem cannot be totally cured with medications. We have outlined already in previous chapters how stress can cause inflammatory responses and how also what stress can be to a cat. There are many environmental enrichment strategies that may be tried, and your veterinarian may dispense medications to help with symptoms, but it is never an easy solution and many cat owners can get discouraged and or get bullied by spouses to get rid of the cat. This may shock you but it happens with such frequency that it makes my head and heart hurt. I have seen spouses nearly split because of a cat recurrently eliminating outside the box. This added stress within the household and the natural retraction or withholding of affection felt toward the feline in these instances actually increases the stress for the cat and makes it even more likely for the behaviour to continue.

Finding a new home for the cat can sometimes help because there can be new distractions, perhaps fewer cats outside in their field of vision that may be upsetting them, or fewer housemates to negotiate with, but often these cats end up being surrendered to humane societies for euthanasia. It is desperately sad. If they are inclined to go outside, then I feel this is a better next step; it is far less deadly and may solve the problem if they will go out. However, the time to start thinking about this is before the problem starts.

Letting your cat outdoors is a viable option when the problem is eliminating out of the box. The idea is concerning for many, but when the options include euthanasia, it is far more viable an option for the cat. After all, it is hardly in their best interest to stay confined if they cannot be their true selves. Let them be their true selves if they are showing you they cannot be so in their current environment. A cat voiding out of a box is saying as much. He is asking for help. If the cat has no urinary tract infection and we have ruled out stones and crystals (which incidentally can also be caused by inflammation) and all we find is recurrence of inflammation, then he does not have the environment he needs and he is not coping. Ultimately, changes need to be made.

Case Study of Boot: Eliminating Outside the Litter Box

Some people are more forgiving than others. I knew a cat some years ago named Boot who took up urinating down the air vents in his home. The smell of urine permeated the entire home. In one respect, it was a thoughtful place to pee, but it was impossible to clean thoroughly. For a long time the owners could not even detect where the smell originated, because Boot was such an expert at his aim down the vents. He didn't even soil the carpet. All manner of treatments were attempted, including isolation and various behavioural medications. When inflammation was found, we prescribed medications for inflammation. Furthermore, all manner of environment enrichment strategies, multiple litter box trials with different litters and styles of boxes, and locations, were put in place. The owners even tried to train the cat on a toilet. Sadly, urinating down the air vents had become a habit that couldn't be broken.

Perhaps if we had seen Boot earlier on we could have done more. For instance, most cats that we treated for urinating out of the box after one or two incidences returned to the box immediately once the cause was determined and treatment began. Even after several weeks of urinating out of the box, most cats can be retrained to the box once the inciting cause is determined and eliminated. But the longer it goes on without treatment and intervention, the lower the chance of a positive outcome. After a year or so of urinating down the air vents at his home, it had become a hard and fast habit for Boot. The family put caps on the vents, which did help with the clean-up, but the cat still urinated just to the side of the capped vents. The family tolerated it. I thought they were saints. They even found ways to cope with absorbent pads and enzymatic cleansers to lift the scent and eventually they had to remove the carpet. Boot may have reformed if they had just let him out, but they lived on a busy street and were worried about outdoor risks. In addition, they couldn't bear to send Boot away to another family. I'm not sure he could have coped in a barn situation, so I was happy they abandoned that idea. This is how some feral communities begin, when their humans give up on them and essentially hand them over to their own care where there is often little human contact. In the end, the owners were content with the reduced frequency of air vent use, which continued to occur when Boot was annoyed with his housemate or a roaming cat outdoors. They followed strict litter box cleaning schedules and put new boxes close to the vents he liked to use. They blocked off one room from him entirely where they felt he might be watching a cat outdoors.

Had we attempted early treatment after just a few days or even weeks, Boot likely would have returned to his box for good. After months and months of inappropriate elimination, the inciting cause can be difficult to sort out. Was it inflammation and pain on urination that caused the first events? A cat that has pain on urination will develop a negative association with his litter box and will seek other areas. Was it uncleanliness of the litter box or insufficient litter box options in a two-cat household? Was it a problem with the location of the box or boxes? Was it a dominance issue where Boot was denied access at times and found his own solution? Was he marking initially? Was it a combination of some of these things?

In most cases it is the location, state of the box, number of boxes issue, or an abrupt change in litter type. Once those litter box issues are corrected satisfactorily, a cat is usually very happy to take to his box again. When it is chiefly an inflammatory problem, then it is more complicated. When there are other health concerns, it is even more complicated. Boot continued to defecate in his box, which isn't surprising in these cases. Even when the litter box issues were addressed and remedied, Boot continued to use the vents because he had developed a preference for them after so much time. After so much time the inciting cause no longer is relevant; the cat has made his choice. Boot's case was more complicated above and beyond the time element than most cases making the prognosis for returning to the box questionable. Boot had heart disease and was on heart medications and a low dose diuretic. This meant that Boot's urine production was greater than average and he had needed to void his bladder more times a day than other cats. This is why I like to use Boot as an example, because like ourselves, cats are complex beings and we can rarely prescribe cookie cutter solutions for problems. Each cat is his own study and each cat requires his own set of guidelines to treat him. A cat like Boot would have been visiting the litter box more often and the litter would have become wet more quickly, which would make litter box conditions poor unless daily or bidaily scooping was done. It meant several treatment restrictions were placed on us in terms of medications we could use. It also meant that the owners were reluctant to impose strict isolation on Boot for purposes of retraining him to the box because they were concerned the stress of it would hasten deterioration of his heart condition. Ideally, a cat is isolated to a small space one week for every month that he has been inappropriately eliminating.

Chapter 7

The idea behind confinement is that a cat will generally not soil his immediate surroundings. So if he is placed in a small area where all his needs are met, like his food, his water, his bed, his toys, and his litter box, then a cat will choose the litter box for elimination. Confining him for the recommended period re-establishes the habit to do so in preference to his new inappropriate site. Boot found confinement stressful and his pleas to be released were very persuasive. The idea of confinement was abandoned after only a few days and so was not effective. Partial confinement may have aided Boot, but we did not try this. For instance, if he had been confined overnight and released after feeding and his first trip to the litter box, this may have been enough. Usually with partial confinement when full confinement does not seem palatable, we also recommend confining the cat whenever the owner is not at home, whenever the owner felt he may have stress (for example when company arrives), and also when bowel movements or urination is expected (some clients became very attune to their cats' schedules by observing them). Usually a cat will settle in to confinement particularly when it does not take him away from his humans. Some even learn to appreciate it if he is living with other cats that are dominant. The object here is not to make the cat miserable or to punish him; it is to retrain him so that he may remain a part of a loving family.

Another helpful tactic that usually helps once confinement is complete is to deny access to areas where the cat previously eliminated inappropriately. When this is not possible, one can try using those areas for feeding and for water bowls, because a cat will generally not soil where he eats. Or, litter boxes may be placed right on top of the areas where elimination was occurring. Gradually these boxes can be moved a foot per week until they are in less conspicuous locations. Other ways to make a location less desirable for eliminating include tin foil, upside-down carpet runner (with the little prickles standing up), a motion detector triggered device that sounds an alarm or mists citronella spray, an electric shock mat (forgive me for suggesting it), pine cones, large marbles, mothballs (these are toxic though) or stones over the area, upside-down mouse traps, two sided sticky tape, or covering the area with plants or furniture. For any cat that has had elimination issues, a few extra litter boxes could save his life. Whatever the case, the areas where the cat had been soiling need to be thoroughly cleansed with enzymatic-type cleansers to totally lift the scent away so it will be unrecognizable to the cat's superior sense of smell.

Case Study of Beau: A Bed Wetter

Beau is another good example of a cat that had elimination out of the box problems. In this case, he was a bed wetter. This cat had been given to a little boy doing this paper route, along with four other kittens, as a tip. Cats have value of course, but they are not currency. It may be argued this was a cruel tip to give a boy who knew his mother would not let him keep the kittens. Can you picture him making his way home with a box full of kittens, his paper bag slung over his shoulders, and his mother's mouth agape in shear shock and disbelief? The kittens were dropped off at our clinic and we found homes for them.

Eventually, Beau found his way back to our clinic as a confirmed bed wetter. His biggest crime that finally got him ousted from his home involved a hockey bag full of gear. He'd probably peed in it all summer and fall before the owner realized, and the gear was hard with dried urine. That was the last straw. Again, it is difficult to know what caused those first events after so much time has passed. Apparently, Beau did not like the dog he lived with, so perhaps getting safely to the litter box was a problem for Beau. Perhaps litter box hygiene was also an issue.

Maybe if Beau had been neutered and then allowed access outdoors he would have reformed. But, he'd been disowned for good. We took him in at my clinic as one of our clinic cats. He thrived there, and he was very affectionate and social. He preferred to be in back with us as we worked, patrolling as though he was keeping us safe. Occasionally, he would go up front to reception and take his turn as greeter. He was known to gently nip a bum or two, looking to turn heads, and this was his way to get praise and attention. He had a long grey tabby coat with a big tufty mane and a white patch at his bib. Because he would not allow grooming, we shaved him and gave him a lion's cut. The first time we did that we found a heart shaped white mark on his chest. He was the king of hearts.

Beau continued to urinate occasionally out of the box, usually in the sink, which we thought was so thoughtful. Sometimes, he peed in one

Chapter 7

of our purses—never mine, I am happy to report. However, this cat did destroy a good many knapsacks and purses, along with all of their contents. Beau liked the smell of his urine; in fact, he didn't like it when we tried to take his litter box to clean it; he would put his foot down in it in an attempt to prevent it. To humour him, we began to leave just a trace of his urine in the box, like a perpetual rum pot, and that appeased him. It was a fine line, because if you left too much urine, then he didn't want to use the litter box at all; he just wanted to have it close by to enjoy his own scent.

I believe Beau would make a lovely attentive affectionate pet for someone with no other pets. Someday I am sure a client will fall in love with him and will want to take him home, but it will have to be someone who no longer has any other pets. If he were a single pet, I believe he would not eliminate out of the box ever again. He only ever does it when he gets upset by a patient, usually the very pretty ones that we couldn't help oowing and ahhing over, or ones that became angry in the clinic setting. He likes to be the centre of attention.

We have had other cats living at the clinic that eventually found permanent homes, so I still hold high hopes for Beau, Chuck, and Coyote. Mommy, a tortoiseshell who had kittens in our clinic, lived with us for years and was a particular favourite of mine. She made friends with a client and eventually went to live with her. She'd been brought to us as a young adult to be euthanized; her owners claimed she was vicious. It was easy to see that aggression was not her problem. She was a confident cat and we all liked her immediately. She was pregnant, but she likely had been treated poorly and acted out in defense, if at all. Pregnancy may have reduced her tolerance for some things, but we never saw any inkling of aggression in her throughout the years that she lived with us at the clinic. She was a terrible mother, but such a dear in every other way. Her real name is Melba but I only ever called her Mommy. One of my lovely clients fell in love with her and asked if she could take her home, and at first I didn't think I could bear it. But Mommy hit the jackpot with this family and I had to admit she was happier there. She frequently came back to us to board and she always made a point of gently telling me that she wasn't staying. I could see it in her eyes that were always just a little brighter when her new family was around.

Marking

Marking is different than urinating out of the box. The intent is different. Or I should say, with marking there is intent. When a cat urinates out of the box due to inflammation or infection, there may be pain. When a cat is suffering from painful urination, a negative association is often made with the litter box. However, with marking, a cat is usually spraying small quantities of urine in an upright position onto a wall or other vertical surface. He is depositing his scent and claiming things as his, and he will continue to do it in order to keep his scent present. It is normal. It is usually an unneutered male action, but some neutered males will do it if they are provoked enough, especially if they were older when they were neutered. It can be helpful to note when and where the spraying occurs in order to give clues as to the cause. I used to ask my clients to draw me a floor plan of their home with windows and doors included and showing the locations (and timing when possible) of spraying events. Usually it is due to an outdoor cat that has crossed some kind of line perceived by the indoor cat. It is a totally normal response in male cats, though totally unacceptable to us humans. In feral populations or with other outdoor male cats, they are able to patrol their territories daily and scent where they feel they need to on a regular basis. It must be frustrating for the indoor cat that cannot make these deposits in a meaningful way to keep those rascals from across the street from frequenting his yard. We upset this balance by confining cats indoors when there are cats outdoors within what they perceive as their realm. We should not blame cats for acting out.

The best form of treatment is immediate neutering if the cat is intact. If he has already been neutered, then your veterinarian might choose to dispense behavioural medications that can help, along with some environmental strategies. Pheromones can help a lot with spraying problems as well and in some cases maybe all that you need. They are available in spray or diffuser formats and can trick the cat into thinking he has made his deposit of scent already. He'll walk by and recognize his scent and think to himself, "That's marked. Good. I can leave it for another day or two." He is just reacting to his environment, unable to get outside to do the marking where he wants to. The behaviour we are correcting with medications is actually an appropriate behaviour. That is what makes it so difficult to manage.

Beau, our clinic cat who liked to urinate in purses, also felt the need to mark from time to time. Thankfully, he did it in the same place every time, and all we needed to do was tape a plastic bag to the wall and wipe it up whenever it occurred. I suspect he did this when he took a particular dislike to one of our patients. He would get annoyed and quietly exit to his preferred place to mark. He was quiet about his disapproval, but certainly made it known. Beau was displaying subtle but effective non-aggression based feline communication.

When Beau first moved in with us at the clinic, the first thing we did was neuter him. When he didn't reform, we tried every medication available for marking and inappropriate elimination. We tried pheromones. We tried Selective Serotonin Reuptake inhibitors and other antidepressants. We even tried hormonal therapy until we became convinced we were risking other problems like diabetes mellitus developing. The thought of more urine that diabetes would promise everywhere and giving poor Beau insulin wasn't exciting to us. All these medications worked for a time but he was devoted to his own scent and didn't see how any of it was a problem. So, we decided we'd need to manage it his way and it really wasn't so bad once we learned to put our purses away and he had found a few places that we could accept to use again and again to continue making his mark either by spraying or peeing. We compromised. He didn't. He had already compromised when he became an indoor cat only.

Case Study of Sage and a Mountain of Siamese: Urinary Tract Disease

Urinary obstruction, also known as Feline Lower Urinary Tract Disease (F.L.U.T.D.) is a common problem among male cats, both indoors and outdoors. It is different than Feline Interstitial Cystitis (F.I.C.), which is seen primarily in indoor cats in that there are crystals or stones found in the bladder and urethra, which cause the obstruction. Obstruction can also occur with F.I.C. but the obstruction in that case is caused by swelling and inflammation around the trigonal area of the bladder, as well as sloughed cells and mucous that aggregate and form a plug that can slip down the urethra and cause an obstruction. F.I.C. can have a crystal formation component, but the crystal formation in this case is usually secondary. In other words, the inflammation comes first, which alters the environment within the bladder and then favours the precipitation of crystals. The crystals then do their work to make the situation even worse and often cause obstruction.

In either case, F.L.U.T.D. or F.I.C., if obstruction is found, then the cat is in extreme discomfort and this is an emergency situation for him. I think the case of Sage, a college mascot of sorts, living in the green house on campus is a nice example of what can happen with F.L.U.T.D.

Sage was a two-year-old beautiful tabby cat, quite overweight, but not obese and he had been dribbling urine for a few days and licking at his penis a lot. He had been eating up until the day before they brought him in, when he had begun to vomit up fluid and be listless. He had been observed to go in and out of the box without producing anything and then he would vomit some more. On his physical exam he was found to be moderately dehydrated and completely obstructed. His bladder was large and turgid and very painful. He cried piteously when palpated.

Blood work showed that his renal values were elevated partially due to dehydration and partially due to the obstruction. He was placed on IV fluids, given pain medication, and anesthesia was induced. Under anesthesia, a urinary catheter was passed with great difficulty because it met two areas of obstruction from tiny stones along the urethra. Once the obstructions were flushed away and the bladder entered, it was emptied and the urine was bloody and gritty. Sage's bladder was lavaged copiously to remove the crystalline material and the catheter was sutured in place for a period of 24 hours.

With such bloody urine, we were very lucky to be able to identify some struvite crystals so we could determine the best treatment plan. Sage was started on pain relief, a urethral relaxer, and antibiotics. He was also immediately switched to a medicated diet that helped to dissolve his stones. After 24 hours, Sage appeared to be doing well; he was keeping his bladder small with the catheter in place and we could see that urine was coming out around the catheter, as well as through it, with gentle expression. The catheter was removed and Sage continued to urinate small frequent amounts for the next 24 hours until he completely blocked again and required a second catheterization. The second time was not quite so difficult and he did well. Sage stayed with us at the clinic for five days. Recheck blood work showed that Sage's kidney values had reverted to normal, so there was no long-term damage done to the kidneys thankfully. We made signs with Sage's picture, medical information, and treatments, and the students posted these around campus. A checklist was made so that he would get his medications on time and that he would not be overmedicated.

Because of Sage's living situation, it was a little bit more difficult to monitor him at home, so he came regularly over the next few weeks to ensure he was okay. He did well. He did have one little hiccup in his recovery a few months later when food ran out and someone bought his old food and he began to eat that again. He started to strain again and lick his penis, but was not fully obstructed. With quick thinking and rapid reaction by the kids and staff at the college, we were able to get him back on track without catheterization. Before Sage's urinary problem was diagnosed, he had been eating a poor quality dry food. He had probably been developing a little sandbox of crystals in his bladder for some time. Eventually, these little crystals began to stick together, slipped down the urethra, and got stuck there.

Poor quality foods can cause crystal formation in specific bladder environments. Dry foods make it worse. We don't know what happens first, whether it is the inflammation that changes the pH environment in the bladder and then the crystals precipitate when poor quality foods are eaten, or if the crystals themselves cause the inflammation and then the obstruction occurs after that. It does not really matter. We know that poor quality food is a big factor because we rarely see this kind of problem with premium quality foods. We know that dry food feeding is also a risk factor because it makes the urine too strong and promotes precipitation of solutes into crystals and also promotes inflammation. Sage unfortunately had never had soft food and wouldn't accept it, but he did well with the veterinary urinary tract dry food that was prescribed.

Sage went through a whole lot in order to be well again. Days of painful urination and then total obstruction, five days in a clinic on IV fluids, two anesthesias and urinary catheterizations, lots of prodding and palpating, blood collection on two occasions, multiple medications, and a change in food. None of this sounds nice, does it? It was all stressful and necessary to get him back to his greenhouse. He was a very brave boy and an excellent patient. It could all have been prevented if he'd been on a better diet and some moist food. It's worth the few extra dollars for good quality food. Basil was very lucky to have so many people looking after him. How were they to know? They couldn't, but don't let that happen to your cat. Sage's students held fundraisers to cover the costs of his care, which is really touching. He is a great cat.

Feline Interstitial Cystitis (F.I.C.) is another common urinary tract disorder we see in both male and female cats. I have mentioned it several times already but just want to tie it in here to be complete. The most common sign of this disorder is urinating out of the box. They may also lick their lower belly because the bladder is painful. The males can obstruct but treatment is usually much easier because obstruction is relieved more easily and once inflammation and pain is treated, they usually reform and go back to the box if treatment is quick and a habit hasn't formed. However, because it is a stress-induced entity, it will recur unless we make environment changes and or keep them on medications to chill them out.

F.I.C. was responsible for many second opinions that I saw in my clinic. I can think of many that would come and you could see that they were thinking that maybe their cat had been incorrectly neutered and that is why he peed out of the box. Others had already been treated for urinary tract problems elsewhere but when they recurred they came to us looking for better treatment that would last and be permanent. Many could not accept that stress could be a component or the cause of the problem. I can think of one particular second opinion that I saw where this was a problem: a friendly enough individual with four Siamese cats that all loved one another, the kind that slept in a mountain of cat. At least one of them (though he could not know for sure that it was not more than one) was urinating out of the box. He had observed one in the act, so we started with him and he did have inflammatory bladder issues or F.I.C. just as the previous vet had identified. The owner was upset when I started asking questions about the size of his home and where and when the urinations were occurring, how many litter boxes, where they were and how often they were cleaned and he was livid when I suggested that there was most likely a crowding problem, a dominance issue, or that there may be some cat outside upsetting his cat. I recommended environment enrichment strategies including more litter boxes at different locations and more regular cleaning and he said I was crazy.

Now we could have extended diagnostics to do ultrasound to look for stones that did not show up on radiographs or some anatomical abnormality, but this was all declined, and I was left to work with what information I had, which was four cats living in a small space with too few litter boxes and one of them is saying something is wrong. The inflammation tells us something is wrong. I think this individual felt like it a personal attack, but really he just wasn't able to accept that cats often need more than what we give them in their little microsphere. Just because we want them to conform to our ideals of what is normal and acceptable, it does not mean that they can or will.

Fortunately, there are newer nutriceuticals, or nutritional supplements on the market that can help the multi-cat household. They are exciting products and I recommend you ask your veterinarian about them if you have a multi-cat household with some urinary tract issues. One in particular, a casein derivative, acts like a calming agent. The calming agent allows inflammation to come under control because the cat no longer feels the stress that creates the inflammation. There is also a new diet developed by a veterinary brand that contains hydrolyzed milk proteins and tryptophan. It has been a blessing to many multi-cat households that have individuals urinating out of the box at times, or even inter-cat aggression problems. In addition, glucosamine chondroitin and other similar products traditionally used for arthritis can play a role in treating inflammation of the bladder as well. They can enhance the protective mucosal barrier lining the bladder wall, which will help reduce inflammation. There is another nutriceutical on the market that contains cranberry, which can help bring the pH of the urine into a more normal range, and this will help prevent crystal formation in some individuals and also help with inflammation. Whether the problem is crystals or just inflammation, the pH will be altered in the bladder and so crystal formation is always a risk. So it is important in individuals prone to inflammation and crystals that their bladder health be closely monitored and maintained. If your cat is urinating out of the box, you need to consult your veterinarian to determine the cause and your veterinarian will decide how best to help. Thankfully there is a lot that can be done. Treatments can vary widely between cases depending on all sorts of factors. A urinalysis and other diagnostics will be needed.

Case Study of Roger: Chronic Interstitial Cystitis

A cat I knew named Roger, who had developed chronic recurrent interstitial cystitis and was urinating out of the box sporadically for years, finally recovered for good when the family across the street moved away with their four cats. So, keep your eyes open. Cats are extremely in tune with what is happening outside. Sometimes closing selected curtains at certain times of the day, or putting a favourite chair or climber in a different window can relieve stress for your cat.

A Litter Box for Each Cat in Your Home and One Extra

Areas for elimination are considered resources to a cat. Litter boxes must be in the right location somewhere quiet where a cat will not be disturbed by loud or sudden noises. Avoid high traffic areas. The basement isn't always the best place, for example, if your furnace is noisy. It is mandatory to have at least one box per cat plus an extra. If you don't want five litter boxes dispersed around your house, then don't have four indoor cats. It is as simple as that. If you have four cats already and have just two boxes, then consider yourself lucky if you haven't had any problems yet. Get yourself a few more boxes.

It doesn't matter if the cats choose to use the same box; you still need the extras. Even cats that get along well will have a social hierarchy. One cat is always dominant and the other cats will always be subordinate. Due to the fact that litter boxes are considered a resource by cats, there will at times be competition for use of these resources. A flick of an ear or a glance that you may not even notice may tell a subordinate cat that he's not allowed near a particular litter box right now. It's confusing because cats don't exhibit their dominance all the time and it is usually very subtle so we don't see it. Sometimes we don't even know for sure which cat is dominant. So, you need multiple boxes in different safe locations. And you need to keep them clean with daily scooping. Cats do not favour scented litter, and this can be a factor that may lead to out-of-box elimination. There are many different kinds of litters on the market now, many of which are made from natural sources like cornhusks or wheat or pine, for example, and are completely biodegradeable. Some of them clump and some do not. Many cats prefer the old-fashioned non-clumping gravel or clay type litters. Others like the softer clumping varieties. The type of litter matters, the type of box matters, the depth of litter matters, the location matters, and the number of boxes matter. In the case of box, litter type, and depth, all cats have preferences. Note: clumping litters may not be the best for kittens or other cats that tend to be messy in the litter box. Clumped material on their paws may cause digestive issues when groomed and ingested.

Remember that the cat chooses, so if you've not given him what he needs, he will choose his own spot to eliminate. Once a cat begins eliminating outside of the box, it may be difficult to change this habit. It is important to remember that the elimination needs of a cat may change as they grow older. A cat crippled by arthritis or one that has chronic renal insufficiency and urinates more frequently, even the obese cat, may not want to go all the way down to the basement any longer. You may need to have all his resources available in a more convenient and smaller space.

"There is more than one way to skin a cat," so they say. Even though this saying alludes to the fact that there is more than one way to solve a problem such as a cat voiding outside his box, I also think that when I hear this saying that a cat is already skinned in many cases when he is confined in a small home with too many other individuals. There is just too much going against him to succeed in this environment and eventually one of the crew will have some kind of problem related to this indoor, multi-cat scenario. And it may result in euthanasia. To avoid this dire consequence, set your cat up for success. If you want your cats to stay in, then follow the guidelines for litter boxes, study environment enrichment strategies, don't overcrowd your cats, and seek help right away when there are accidents outside the litter box. Think outside the box. Think like a cat.

Chapter 8:
All Cats Are Grey in the Dark

The cliché, "All cats are grey in the dark," suggests that the physical appearance of a cat is not that important. There are other concerns for the cat that are far more important than its outward appearance. Your cat does not spend an hour in front of the mirror each morning admiring him or herself while getting ready. He knows he is beautiful. This is one of my least favorite cat quotations because it actually refers to women. It is understood that all cats are beautiful, but are still grey in the dark. The same apparently is not true of all women, but with the use of this quotation, men can pretend their lovers are beautiful in the dark. The saying originally appeared in John Heywood's book of proverbs in 1546, where no doubt it was more innocently meant that true beauty lies within. Cats have superior vision in the dark as well as other heightened senses. I will also discuss the issues in cats that we often don't want to talk about, the ones that we'd rather not think about, and rather keep in the dark.

The Senses

This quotation reminds me of something very special about felines, which is their ability to see well in the dark. This is due to the unique anatomy of their eyes, which are proportionally larger than many other species' eyes, and so the area of their lens and cornea is larger and allows more light in. Their peripheral vision and angle of vision is very good but they do have a small blind spot between the eyes, which is why often they can see the water bowl clearly from a few feet but as they approach it, some use a paw to confirm that it actually holds the clear fresh water that they watched you put into it. When dilated, their slit pupils help to gather more light. They have more rods within their light-sensitive retinal tissue at the back of the eye and their retinal cones are extra sensitive to light. In addition to all this, they have a reflective layer at the back of the eye called the tapetum lucidum that reflects light back to the retinal tissue intensifying it. It is the tapetum that gives that glow in the dark appearance like two yellow green flashlights when light hits the eyes. Cats can see in the dark with one-sixth the amount of light that humans need. Cats do see colour, though not as well as the human eye. In the dark, not every cat will appear grey to a cat. Although they are near sighted, they see movement very well. These eyes are adaptive to their natural nocturnal habits and hunting.

In addition to their excellent night vision, cats have an excellent sense of smell thought to be 14 times more sensitive than our own. Already well developed at birth, it helps the young find the mammaries, and is important their entire life for inter-cat communication. For instance, Watson loves to show me his bottom, tail up, right in my face when he is being affectionate. I'm not so keen on this display of his scent and other attributes under the tail, but he seems to offer it as a gift. A fellow cat would understand this differently no doubt. Face-to-rump is an important mode of identification in cats, following face-to-face. They have multiple means of depositing their scent for communication. Through their cheek glands (pheromones), saliva, which they deposit when they rub their faces, anal glands (deposited on their stools), their paws when they scratch, and of course their urine, which they spray usually on vertical surfaces when standing upright. When we open that can of tuna, it is not just the sound that brings them running, but it is very likely that they have already smelled it. They are also able to detect sounds in the range of tiny mouse squeaks, much higher in frequency than our ears can detect. Their earflaps help locate, funnel, and magnify sound frequencies. In fact, their well-developed sense of hearing is much more important than their sense of smell in hunting. Taste is more difficult to study but it does appear that their taste buds respond well to sour, bitter, and salt, but minimally to sweet. The sense of touch also appears important to the cat. A mother's touch is soothing for the newborn. At just a few weeks of age, they have a well-developed rooting response and will push their heads into warmth. As adults, they prefer gentle stroking to the harder patting that dogs often enjoy and generally less restraint is better than more for the cat in a veterinary setting or at home. Most cats don't appreciate a lot of petting near the base of the tail. Though, a little is good. There is always a fine line with a cat. I like that they know what they enjoy and don't enjoy, and aren't afraid to let you know.

Chapter 8

Combine the above senses with their sensitive whiskers, their efficient teeth and jaws, stabilizing tail, amazing agility, and athletic abilities, and you have the perfect predator.

I wonder what it is that a cat sees when he looks upon a human. Like humans, they combine all their senses to understand what they see. The cat's senses are not like ours—they are superior for the most part, and their perception of their surroundings and those sharing the surroundings must also be different from our own perceptions. Taken individually, we can understand each of these heightened senses but if we put them together, then the cat's experience of life isn't something we can fully appreciate. We humans must often seem to be very loud and smelly to a cat. A house full of humans may very well take their breath away the way a stinky litter box does to us as we enter the house after time away. What intrigues me most is the idea that the cat has that sixth sense: the ability to see and feel pure energy. Some humans claim to be able to do this, like see auras, for instance. Anything sense wise that humans can do, a cat seems to do better, so perhaps all or most cats are able to do it. Maybe a spooked cat really has seen a ghost. A cat is always an optimist and knows none of us are lost causes. Perhaps a cat understands these things because he can actually see a shift in a human's aura as he plies his affections. That cloud that clings to us when we are not our best selves—maybe the cat sees strands or filaments of light there too. He takes his time shining us, removing all the smudges so that, like the prisms that we are, we can emit our true colours. This is my theory, anyway. This may sound like crazy talk to many people. I am stretching different mental muscles now that I do not need to be so deeply grounded in my left-brain. My left-brain of course wants more proof, but my right-brain does not need it. It is already convinced of a cat's sixth sense.

A cat doesn't hide this skill, but we may not see it unless we are watching closely. It goes along with his superior intuitive intelligence. We can ask them what they see in our futures, but they will stay mute on that unfortunately. There are other things as well that a cat does prefer to keep in the dark.

Hairballs

A friend of mine sent me a link regarding marketing ideas for products made from hairballs. I was going to include it here but I knew the cat would disapprove as it is very undignified. Christmas decorations, purses, necklaces, clothing, that sort of thing. This friend found her Persian cat's frequent hairballs quite impressive and over time her family began to joke about what they could do with them. A cat is less impressed by hairballs. I certainly feel the same since every time I cough someone asks me if I have a hairball. Oddly, this tiny tease has caused me to have a great deal of sympathy for this feline phenomenon. I think instead cats are disgusted by these demons that overtake them from time to time. They immediately disown their hair in these monstrous forms as soon as it is emitted. My technician and I used to make similar remarks about what one could do with cats' matted coats that we had shaved away. Some are like the full sheering of a lamb. Now there is a business idea!

From February through July, if I saw a cat off his food or a vomiting cat, then the diagnosis was hairballs until proven otherwise. Cats are fastidious in their grooming habits, so they will have hairballs from time to time. If it occurs more than every few weeks or so, there may be problems with gastrointestinal motility at fault, or your feline may have some kind of skin condition that causes over-grooming or excessive loss of hair. Hairballs are not always normal in a cat. It's very important that you seek your vet's advice if your cat has frequent hairballs, especially if there is any change in behaviour or appetite.

Hairballs can occur regularly or irregularly throughout your cat's life. Certainly there is a hairball season and that begins as soon as the days become appreciably longer in late winter. Cats of any age may become very ill from their own hair and it can appear like a life-threatening ailment with profuse vomiting, anorexia, and dehydration. Many cats will require hospitalization and intravenous fluids as well as help passing the hair. Any healthy cat with a healthy dose of self-esteem will insist on keeping his coat in place and in good order. In addition, having a friend he likes to groom is double the trouble, especially since cats shed more during hairball season. With multiple cats in a home, it is not always possible to tell who is vomiting by the colour of the hair that is regurgitated.

Chapter 8

As I write this, I am fully aware that there are no less than three little mounds of hair vomit in our basement waiting for me to clean them up. A few look like tubes of solid hair. Other times they may look like mostly liquid with bits of hair. I stepped in another one on the way to the bathroom overnight. It's time to get the hairball remedy out and treat them each with a healthy dose daily for the next few days. I am not worried because the cats all are eating well and are active and playing and are still insisting on grooming. But, they need some help and I watch them closely. Hairballs can cause serious illness, so it is best to visit your vet before things get too out of hand.

When a cat is desperately ill from gastrointestinal obstruction and irritation from hair, it doesn't matter what the cause is. You still need to treat it. Once a cat becomes dehydrated and is no longer eating, he isn't going to get better on his own. Felines need to be rehydrated and they need treatments to loosen and lubricate the hair. With veterinary treatment, eventually they will usually have success either passing the hairball, or obstruction in the stool, or vomiting it up in smaller pieces. Sometimes large hair masses causing obstruction and illness need to be removed surgically. After some years in practice, I became amused when owners would call to report, after we had treated their cat, that the hairball had finally been brought up, and what a waste of money that treatment had been. Hairballs are a legitimate illness and they can be deadly.

It is important to understand that some cats with hairball gastritis may not have evidence of hair in their vomit. I treated a cat many years ago for chronic gastritis. We had her on special diets and anti-inflammatories and stomach settling agents, and she just kept on vomiting periodically every few days to every few weeks. Finally, we gave her an endoscope exam and there in the pit of her stomach was an enormous golf-ball sized hairball. The owner had never seen any trace of hair in her vomitus, but the cause of her chronic vomiting was most certainly this enormous hairball.

Groom your cats regularly to keep the ingestion of loose hairs to a minimum and please use hairball remedies through late winter to early summer to help them pass hair without incident. As well, many modern diets have hairball prevention elements within them that help with passage of hair.

The Unseen Parasite

Most of you have not had the experience of looking through a microscope day in and day out so you cannot see or possibly imagine what creatures lurk around at this level. Although we typically cannot see them with the naked eye, internal parasites deserve special notice. Parasites are very "clever" little creatures. They keep their house clean. They take only what they need. They don't upset balance. They don't want to have to move, and so they find their own balance between feeding on their host for propagation of their kind and not overfeeding to the detriment of their edible home. Most healthy adult cats can harbour a little city of parasites in their gastrointestinal tract without showing any symptoms whatsoever. A cat will not vomit up a worm until he is infested with many and that is the only way you would see them.

When a cat is under the weather for some other reason, the parasite load may overwhelm him at that point. A long-standing parasite load will eventually cause weight loss, poor coat quality, and possibly even illness including anemia. Kittens, young adults, and senior cats can become very ill from parasite loads, and often fatally ill.

Just because you cannot see them, it does not mean they are not there. Most kittens are a product of cats that have been outside having sex and getting pregnant. Most cats that go outside and have sex and get pregnant also hunt and eat their prey, dig in soil and eat bugs, which are all sources of parasites. Immature forms of some parasites can be transmitted through the mother's milk, so most kittens have a city of parasites in the making.

We had jars and jars of preserved parasites in our clinic, but even I didn't like having them on display in the exam rooms. Most clients found them very upsetting, so we kept them in a cupboard and brought them out only occasionally. Perhaps I was wrong about not displaying them though. Maybe I should have kept them where they would be seen, because they are part of our cats' world after all. In the short of it, we just can't see them unless they are collected during an autopsy and are placed in a jar.

Fecal exams are very helpful for detecting some parasites, like the single cell parasites like giardia and coccidia. These fecal exams however are notoriously bad at detecting roundworms and tapeworms, the most common parasites seen in cats. We are looking for the microscopic eggs, not the worms, because the worms are holding on tight to the gastrointestinal mucosa and aren't going anywhere. The roundworm eggs are shed cyclically and so a single fecal sample brought to a clinic will often be negative for roundworms. Likewise, tapeworm eggs are shed as little packets of eggs in a segment, so unless we can identify a segment grossly in the stool or attached to the hairs around the anus, we cannot detect them because we will usually not see the eggs microscopically. Tapeworm segments detach from the larger worm once they are mature and make their way down the ailimentary tract (intestines) and out of the anus. They can appear singly or in multiples at the anus or in the stool. When they ooch out of the rectum they can cause some itchiness and the cat may groom excessively to relieve that. The segments dry quickly there attached to the hairs around the anus. They can resemble rice or little sesame seeds once they are dry and will eventually drop off into the environment to be picked up by another host such as another cat, a mouse, or even a human.

Kittens need deworming for roundworms at a minimum and tapeworms if the segments are seen(or risk is felt to be high as in flea infestations) and for any single cell parasites found on fecal exams. Several doses of deworming maybe required on a timed schedule depending on the agent used, because single dosing generally only kills adults—not the worms that are in the process of developing, or the forms that have formed cysts within the body outside of the digestive tract. Some intestinal parasites in their larval or immature form do migrate into tissues such as muscle and liver and form cysts there.

The problem about not taking notice of what we cannot see is that parasites shed eggs into their environment. They will be in your cat's stools. They will be on his coat as he grooms them away from his anus and then onto his face, body, and limbs. They will be on our hands when we cuddle our cats and then in our mouths when we casually put our hands to or in our mouths. Some parasites can be very dangerous for humans, especially children, elderly, or immunocompromised individuals. Other parasites of concern include lungworm, liver flukes, heartworm, and various blood parasites.

I hope that these descriptions give you a visual of what I saw when I saw a hunter or a kitten come into my clinic. There are routine deworming guidelines to follow. It is a good idea to follow your vet's advice on this. Deworming is not a cash grab for vets; it is important, and that is why vets recommend it. It makes me want to cry when deworming is declined. I always feel I haven't done my job well when that happens, and it does occasionally. For adult cats that go outside, deworming every three months should be adequate to keep the internal parasites under control within your cat's body and the environment.

A parasite is just as the word says: it lives unseen, exploits the host, and gives nothing in return, except misery. There is one exception to this rule and that is a single celled protozoan parasite named Toxoplasma. This parasite can infect any mammal including humans but the cat is the primary host of this organism. This means that the cat is the only host in which the parasite can produce eggs for dispersal. In other hosts such as mice and other rodents, deer, sheep, and pigs, the parasite forms cysts in the muscle and can only be transmitted to others if their flesh is eaten. Once the cat is infected, usually through ingestion of prey like rodents, it will only shed eggs in the stool for a few days up to two weeks, and then it will stop. Dispersal of the eggs to other hosts such as mice, deer, and pigs occurs through oral contact with the soil and contaminated water. The eggs are not immediately infective and take up to five days to become a danger to another host. What is interesting about this parasite is it can affect the brain of mice and other rodents in such a way that they no longer have an aversion for the scent of cat. This makes them easier prey for the cat, which on the surface facilitates predation and would therefore appear to help the cat. Its primary goal is its own far and wide dispersal. Indoor cats are unlikely to become infected because they are less likely to hunt. However, if they are fed raw or poorly cooked infected meat, then they can contract the parasite through the meat.

Humans may also become infected through hand-to-mouth contact with the cat's stool after cleaning the litter box, playing in a sandbox, or digging in the garden. However, it is felt that the most common source of infection for humans is through ingestion of poorly cooked meat sources such as lamb, venison, or pork. In most cats and humans, there may be zero-to-mild flu-like symptoms only, but in some individuals, especially those that may already be immunocompromised, it may cause serious disease such as encephalitis or uveitis (infection within the eye). It is of particular concern in pregnant woman because it can be transmitted transplacentally (across the placenta), which causes serious problems for the embryo and infant. This is why it is not recommended for pregnant women to clean litter boxes of cats. It does take up to five days for the eggs to become infective, so if the litter box is scooped daily then it is unlikely to be a problem, but for safety's sake it makes sense to have someone else in the family to scoop the litter. Wikipedia says up to one third of humans have been infected with this parasite worldwide, but it rarely causes problems. It is felt that it is much more likely for humans to contract this parasite from eating poorly washed vegetables from the garden or from eating poorly cooked infected meat (usually the latter) than from their cats.

All in all, parasites of all descriptions are nasty unseen things and all precautions should be made to limit their numbers in our cats, in our gardens, and in every other place. Personally, I loathe internal parasites because they are so dishonest and they don't show themselves, which I find rather cowardly. I can work with a flea, however, because I know what I've got with a flea.

Nutrition

I'm a food snob. I admit it. But, there is cause, when the topic is feline nutrition. The cat is also a food snob, as they are widely known to be finicky. A certain food can taste divine one day, and the next he may look at you insulted as though you've served him chicken feed. Some cats make quite a practice and show of declining their food, by holding out for a certain taste, shape, texture, temperature, odour, or freshness. And he's sure you have it right there in the cupboard, or fridge, and he will seem to say, "Just give it to me already!" It can be a lot of work to please the feline palate, and I've known clients to make special trips to the store to try to please their cat, sometimes daily. Cats are often reverse snobs in that they love junk food. I've frequently wondered if certain brands that are widely popular among cats might contain traces of opium or maybe catnip in them to get them hooked. In fact, many modern cat foods taste so good that many modern cats have lost the taste for their ancestral food, the mouse. Perhaps if it were stuffed with shrimp and salmon, roasted in butter, and then pureed, canned or baked, pelleted, and then sprayed with more tasty fat, then the mouse would be appealing again. But for others, the taste of fresh kill and blood still rules.

Nutrition matters, and not all foods claiming to be premium quality are good nutrition. This is a fact. We are starving our cats with many mainstream feline foods. I'm not trying to sell you anything here; I just want you to know that the food you choose will choose many of the problems your cat will have in his lifetime. For some reason, the nutrition topic does seem a little charged and that is why I include it here with other in the dark topics.

So, how do you know what is good? You don't. The label can't tell you. The commercials don't tell you. Your cat won't tell you because he likes junk food as much as I do. His coat may eventually tell you if some other health issue doesn't first. Unfortunately, that may take years, and in the meantime he may have developed other problems associated with poor nutrition. Generally speaking, the more expensive brands will have better quality meat proteins to build and rebuild their bodies, but even this is not always true. I've been so disappointed and enraged by feline food products over these many years. Cats are obligate carnivores. They are not built to tolerate large quantities of carbohydrates from plant sources. Their digestive tracts are shorter than omnivores and they lack the enzymes needed to digest carbohydrates efficiently. They are built to be meat eaters. There are sets of specific amino acids, specifically arginine and taurine, that have to come from their diet, and these can only be found in certain meat proteins. Not all proteins are created equally. In veterinary school, the point was made by explaining that leather is protein, but has absolutely no nutritive value what so ever. And that is why checking the label for protein content cannot tell you anything. There is no other way to say it. Quality matters. Large sums of money spent on advertising and packaging doesn't. I wish they would just spend the money on the product instead. Meat sources are more expensive than plant sources and that is why good quality cat food is expensive compared to poor quality cat food and dog foods.

Chapter 8

The most normal food for a cat would probably be a mouse. Ten or so mice a day would meet all his nutritional needs, minus the parasites, but who in their right mind is going to can a dozen mice? I suppose intensive mouse farming for the purposes of feeding our cats isn't any crazier than the intensive pig, beef cattle, and poultry farming we already do to feed ourselves. Personally, the idea of mouse farming suddenly makes the other forms of intense farming seem just as crazy.

Our modern day diets have come so far away from what is normal for a cat that it is a wonder they do as well as they do. I see some cats do well in fact on the worst possible diets, but that is the exception, not the rule. And perhaps these cats that I have marveled upon actually have been supplementing daily with a mouse or two. Or perhaps they have super-duper genetic makeups already evolved to handle poor diets. Or maybe it's a bit of both. Eventually, what we see with mainstream popular diets is bladder, urethral, and kidney stones; urinary obstruction or blocking; obesity and diabetes mellitus; allergic disease; skin issues; aggression and hyperactivity; and immunity issues, to name a few.

The two things that scare me the most about mainstream cat diets are obesity and the associated eventuality of diabetes mellitus, and urinary tract obstruction. It is one of the saddest things to see when a cat is in so much pain because he cannot void, his bladder huge and painful, his kidney values soaring within his bloodstream, which make his head also hurt and his muscles ache. He will be nauseous and he will be lethargic. He will be dying before your eyes unless he has immediate treatment. He will be dying largely due to the food he eats. Disease processes are generally multifactorial, but diet plays a huge role here.

Dry food is a mistake, for the most part. We made a mistake by introducing dry foods to cats. It is the same mistake we make when allowing sugary cereals, pop, and potato chips into our own cupboards. It tastes good and we like it, but it's bad for us. Dry food was originally formulated for dogs, specifically northern breeds of dogs running sleds, so the food would not freeze solid. The convenience and no mess appealed to us humans because we are a breed that very highly values convenience and no mess. It quickly and forever spilled over into the feline food industry and not surprisingly, we found that there were ways to make it delicious for the cat—even more delicious than a mouse. For years, we patted ourselves on the back because we thought we were doing something good for them and that it was a way to keep the cats' teeth clean, but it didn't keep their teeth clean. Anyone who has a cat that occasionally regurgitates knows that most of those little kibble are mostly swallowed whole. Over the years, I have not been able to say that cats that eat dry food only have significantly better teeth. I observe they accumulate tartar perhaps a little more slowly, but comparing two eight-year-old cats, one eating soft, the other dry, there will be little difference, because both need dental care. I do notice outdoor cats have better teeth, or at least less tartar, gingivitis, and periodontal disease. We do now have some very good effective dental diets, which are larger than your average crunchy and coated, with enzymes to break down tartar.

The main problem about dry food is this: it's not wet. Certainly a cat eating a diet with only dry food will drink more in an attempt to compensate, but it can be difficult to drink enough to compensate. SO this leads to overly concentrated urine, which is far more concentrated than on a soft food diet. This can lead to, or at least be a factor, in urinary tract issues including crystal formation and inflammation. In older cats, I am also convinced that it will hasten kidney changes, although I do not believe it is the cause of kidney changes. Simply, dry food is wet food with extra binding agents in the form of carbohydrates from plants, so that it can be extruded and baked. This is an additional problem for an obligate carnivore unused by nature to handle so much carbohydrate. When I meet a new diabetic cat, it is usually a dry food eater.

If your cat is a dry food eater, then look at ways of increasing his water intake. Keep the bowls clean and his water fresh. Some cats like very cold water so you may wish to add ice cubes periodically or just freshen it frequently. Cats that like a moving source of water like water taps will favour water fountains. Add a little chicken broth or tuna juice to increase consumption. Play with bowls to see what attracts your cat most. Some prefer to drink and eat from low flat bowls like pie plates, or larger bowls, so that their whiskers do not hit the sides. If your cat favours toilets and tubs, then this may be the case. Others like to drink from a cup or glass.

Be wary of any food claiming to be multistage. Yes, a mouse could be considered to be multistage nutrition, but when we take nutrition into our own hands for our cats, it is difficult to get the formula just right for all stages. A kitten needs more protein and calories than an adult. A senior cat also has very specific needs and does better when the phosphorous is restricted a bit and there is less acidification of the urine created by the diet.

I would never buy a food that regularly goes on sale. If it is made with consistent healthful ingredients and is fairly priced to begin with, then it should never be on sale. Unfortunately, many diets are not consistently made and many are not made with mindfulness. It is a way to make a very big margin of profit from inexpensive things that almost look like food. It is the one thing that used to make me very angry in my work as a veterinarian, because as the consumer you are just trying to do your best, and how can you do that when there is so little out there that can be counted as good nutrition among the mountains of pretty products?

My advice is to not get hung up on labels, because they typically don't tell you everything you need to know. Don't get hung up on ash content or any other thing printed on the bag. I don't care if it claims to be all natural or organic or that meat comes first or whatever it says; there are not enough regulations on packaging to ensure that these claims are always true, and yet the claims make it okay to demand a higher price. It can depend on how ingredients are weighed or measured (by dry weight or wet weight, has the meat been injected with water first, for example) or if they have been pretreated in some way to add or subtract weight. What I care about is the result. Ask your veterinarian which diets produce the best results. They have the experience to know which diets work and which ones cause repeated health problems seen in veterinary practice. Start with your vet, then research the brands he or she recommends to make your choice.

I believe all cats should have at least half of their daily food in the form of soft foods. As they get older, soft becomes even more important, but acceptance will be poor if they have not had it in the past. Just like a human toddler, kittens need to experience and accept food textures in order to be open to having a soft food diet. If you feel inclined to make your cats' foods, it is difficult to do. It must be carefully balanced with vitamins, minerals, and essential amino acids that the feline is unable to make for himself.

I am also wary of raw food diets. I am too worried about bacterial contamination to support them. We run into those kinds of problems even in food meant for human consumption where there are strict regulations for cleansing, sterilization of equipment and, strict inspection guidelines for all meat sources. This just isn't in place for pet food and so I cannot recommend it. There is too much risk of Salmonella, E. coli, and other poisonings.

Food allergies and sensitivities are common in cats. Symptoms may include vomiting or soft or runny stool but very often there are no or minimal digestive signs. Instead what you may see is skin problems. There may be a generalized redness and or itchiness to the skin. There may be skin lesions or sores. The cat may flick his ears or hold them down due to inflammation within the ear canals. Anal gland inflammation may also occur, which will cause the cat to groom his bottom excessively or as close to it as he can get if he is overweight and unable groom there. He may also scoot his bottom on the rugs to relieve the discomfort. Food allergies are usually due to a protein. It is thought that beef and fish are the two most common allergens for cats with food allergies but they can be allergic to any ingredient including plant additives or dyes. Sometimes we are lucky and can just switch a diet and relieve the signs without knowing what the ingredient was that caused a problem, but often it takes a while to find a diet that will work for an individual.

Hypoallergenic diets have appeared on the market due to the frequency of food allergies among pets. These diets typically contain few ingredients and have novel proteins such as duck, lamb, or venison. We call these diets elimination diets because they eliminate all common food allergies and help get to the answer more quickly. Food trials on these hypoallergenic diets must be at least 12 weeks long because we know it takes that long for the allergens causing skin reactions to come fully out of the skin and for the skin to heal. During this time, no other food or treats can be given. This is because a tiny little bit of an allergen will cause the reaction to persist. It is like peanut allergies in kids. Even a whiff of peanuts in some individuals will cause a reaction. The same is true for food allergies in cats, except that the reaction typically shows up as continued skin problems instead of the anaphylaxis we see in kids with peanut allergies. Blood tests and skin tests exist to determine what allergies are present in an individual cat and may be recommended by your veterinarian depending on the severity of signs and other factors. It is important to note that not all diets claiming to be hypoallergenic are truly hypoallergenic. Those found in veterinary clinics will be because they are made using equipment dedicated to their production only. Those found in pet food stores often are not truly hypoallergenic because they often have been produced in facilities that use the same equipment to make other diets. This means that they may have traces of other diets within them. Even a trace will cause a reaction, so with true allergies, you are not any further ahead. What I recommend when a food allergy is suspected is to commit to three months on a veterinary hypoallergenic diet and see what happens.

Many popular mainstream feline diets equate to hotdogs, French fries, potato chips, and pop every day for every meal for your cat. Of course they will love it—so would your kids, but you wouldn't feed your kids like that. Many clients bemoaned that they could not get their cats to eat anything else and so what were they to do? It is true that the cat decides in the end. So, don't ever start with these diets. Ask your vet, get him on the best possible diet as a kitten, and adjust the diet as he or she goes through the life stages. I still feel it is possible for most cats to adjust to better nutrition if you keep introducing it a little bit at a time. They will have more energy, they will consume less, they will have better coats, and their stools will be less smelly and less voluminous. They will also be more likely to reach their potential. In the end, they will live longer and in better health and hopefully with fewer visits to the vet.

I want your cat to live a long and healthy life. Good nutrition goes a long way to get him there. Please ask your veterinarian. They know better than anyone else what diets can be depended upon. They see the results day after day. If a vet has just unblocked half a dozen cats eating a particular diet in the past month, or you've seen multiple skin issues or gastrointestinal sensitivities, or frequent recalls of a certain diet, then it will be difficult for the veterinarian to encourage that diet. If it's cheap, the cat likes it, and it makes their coat shiny, and he seems to be thriving on it, then the food salesman will continue to sell it to you. And we will see you later, in a few weeks, in a few months or a few years as other risk factors come into play, with your cat's bladder full of sand or even stones, obstructed, for example. I am not saying that pet food store employees do not have the best of intentions; I think they do. An exceptionally knowledgeable and concerned individual owns our local pet food store, and he does a marvelous job directing his clients to good nutrition. However, many are simply good salesmen, and though may have good intent, they just don't have the same experience as a veterinarian, so their focus and their points of reference will be different.

Pet food stores most certainly do carry good quality foods, but to cater to every economic stratum they must also do their best to carry alternative lower cost and lower quality foods as well. With so many different foods side by side and the labels showing little difference between them, it is confusing for most pet owners to make a decision. In the end, price too often becomes the deciding point, simply because we do not know what the difference is and we do not want to overpay. I know many people feel concerned when they read on pet food labels, "poultry meal" or "byproduct meal" or just plain "chicken byproduct," for example, as ingredients, but truly in many cases this is responsible use of bits and pieces of meat and other tissues that may not seem as palatable to humans but is still very good nutrition for our pets. This should not be a deciding factor either. To my mind it is excellent use of a life that was taken for our purposes, to use as much of it as we can makes it less wasteful and more meaningful. Words on labels such as whole chicken are not necessarily any better than chicken byproduct meal. It is not the entire chicken that they are using, though many manufacturers hope we will assume this. It means a certain percentage of the carcass has been used, and usually it means the head, neck, backbone, and feet. Pet food labels are another kind of secret language. This language contains many dirty little secrets and as soon as you think you understand it, it changes. Proteins are not equal, but in some digests and meat byproduct meals, there will still be good proteins there. In other words, what you cannot know from the labels, you will eventually know from the results: your cat's health. But, don't wait for that. Ask your veterinarian what produces the best results. And remember to always transition slowly to a new diet so that your cat's digestive system will accept it better.

Chapter 8

Remember, veterinary clinics can stock any food they want. They can purchase any food from wholesalers, just like the pet food stores can, but they don't. They carry the foods they trust. Veterinary brand foods will certainly be better nutrition. They put the money into product development, research, and the product themselves. And then they let the results do the advertising for them. A less smelly stool due to absence of certain indigestible solids used in mainstream diets is one of the first things clients notice. These indigestible ingredients in many dry foods alter the bacterial microflora of the cat's digestive system and ferment within the digestive tract creating smelly stools. But this is only one of the many important ways veterinary products differ from mainstream cat food products. Their products are made with mindfulness and integrity. They have a secure and steady supply chain. They are devoted to the idea of nutrition, which is actually becoming a rare thing in the world of pet food.

Veterinary food brands have amazing maintenance diets for your cats, but they also have an extremely thoughtful line of therapeutic diets that work. Louis benefits from a joint mobility enhancing diet and does not need anything further in the way of medications to treat his bad hips, and it also helps his colitis problems. He is drug free. Watson, who loves to eat, is on a formula that satisfies without overeating. He no longer nips my ankles between meals and he is losing some weight. Other formulas developed by these brands exist to treat or aid in treatment of nearly every disease process or health problem you can think of. I prescribed these diets regularly in practice because I knew they worked, tasted good, and were fairly priced. I am in love with these diets because they made things so much easier for my patients and me. Raising health through nutrition makes very good sense.

Good food costs money, but I have learned that it is all relative. For some, a certain diet may seem very expensive and they will be proud to tell you what it costs and what the label says and they will be incensed when you tell them it may cause problems. And then others will feel that same diet is not at all expensive and they will be more open to the idea that they could do better for their cat. Basically, a good rule of thumb is to decide what your budget for cat food is and then go out and find the most expensive food that will fit your budget, but ask your vet for suggestions. Veterinarians love to discuss nutrition, because they know it is one of the areas of their expertise where they can be most impactful on your cat's health. If they don't seem engaged, then you need to find a new vet. Any well-meaning vet should have some passion about nutrition; it is the building blocks to good health and longevity.

Following a budget is the best way to ensure you are doing the best you can do for your cat. After all, we can only do our best. Then, visit your pet food store and make your selection. And make sure at least half of his food is in the form of a canned soft food. At least you are feeding a small cat and not a big dog. Budgeting too rigidly on your cat's nutrition may be false economy, because you could very well be spending large sums at your veterinary clinic when things go wrong.

Remember to measure your cat's food. Free choice feeding can work for some individuals who are not focused on feeding as entertainment, but for many multi-cat households, one cat often gets more than their share. For those cats, the food should be measured and divided into meals. Sometimes it is necessary to physically separate cats for feeding so everyone gets what they need and no more or less. An active adult cat needs about 50 kilocalories per kilogram of lean weight daily. A kitten will need more; a sedentary indoor cat will need less. For an average 10-pound adult cat that means 250 Kcal per day. Many diets on the market do not give you a kilocalorie count on their labels. Sometimes you can find that information online but if you cannot, I recommend you feed another diet. This information is important and should be readily available. In fact, it is really the only thing on a label that can be helpful to you.

The cat's love for catnip is an entertaining thing to watch. Catnip belongs to the mint family and contains a substance called nepetalactone, which is a mild hallucinogen in cats. The cat senses this substance through receptors in their vomeronasal organ at the back of their nasal cavity. Kittens tend not to react to catnip. Some adult cats are not too interested in it, but others have quite strong reactions to it. Very few cats may even become aggressive when exposed to it. Some cats will sniff it; others will eat it. Many will roll and roll on it as if in ecstasy and can appear quite loopy or intoxicated. The effect wears off almost as soon as they leave it. It is not at all dangerous, and if your cat enjoys it, there is no harm in giving it regularly. It can be helpful to train a cat to a scratching post or to distract him when he is a little stressed. A client of mine made pesto from catnip she found in the garden of her new home thinking it was basil. It was not the flavour she was expecting, but she had no ill effects and no high. Like other mint leaves, catnip leaves can be made into a tea for humans that can relieve an upset stomach, insomnia, and can be slightly sedative in nature, but they cannot be used when pregnant. It has other old world medicinal uses when used as a poultice.

Honeysuckle can provide a similar high in some cats. The berries are toxic to cats but the woody part of the plant can be shredded and placed in little pillows and toys and many cats will have similar reactions to it as they do catnip.

Cats often enjoy eating grass and this usually not harmful. Why they do it isn't entirely understood but there are some benefits. One is that it is a natural laxative and so may help with hairball expulsion or even mild stomach upset. It contains folic acid and so cats can benefit from eating grass occasionally. Most often they chew on it and shortly after vomit it up again. Make sure it has not been sprayed with chemicals. You can buy and grow cat grass in your home and there is no reason to worry about your cat enjoying it. However, stiffer grasses or other vegetation found outdoors can cause some issues with obstruction and irritation of the gastrointestinal tract when consumed.

Live Food: Look What the Cat Dragged In

If your cat were choosing a restaurant, it would definitely serve live food. Cats will hunt; it is one of their raisons d'etre. Prey is a good source of protein and other nutrients; it is perfectly balanced and even helps your cat keep his teeth clean. Prey species that cats would be interested in are generally vegetarian and so the cat obtains the small amount of plant material that he needs already partially digested within the preys digestive system. Mice and other prey do carry parasites, however, and so if you have a hunter you will need to deworm him under veterinary supervision every three months to keep him parasite free. Over the shelf preparations sadly won't do the trick. If you doubled the dose of some of these products, they still wouldn't work and they would hurt your cat. Veterinary prescribed deworming agents work and are safe. Remember many parasites that cats can carry can be passed on to humans and can be dangerous, especially for the young and old among us.

I know some well-meaning cat anti-enthusiasts feel cats quickly decimate the songbird population and for that reason should be kept indoors. Well, they do catch the odd bird, it is true but we do far more to decimate the songbird population by our destruction of their habitats by building our monster homes and shopping malls. Put a bell on your cat if he is catching birds at your feeder. Cats do such a wonderful job keeping the rodent population down (your neighbour forgot to thank him for that), that the odd bird here and there must be forgiven. I am an avid birdwatcher myself and so I too would be pretty upset if one of my cats ate an Eastern Bluebird or even a house sparrow innocently perching for a few seconds to feed from my bird feeders. I am not trying to be flippant, but it is the nature of things. It is possible that hunting cats make an impact on prey species for other predators, like foxes, hawks, owls, and coyotes, but again this is natural. There will always be cats that hunt. However, this is yet another reason to practice feline population control by spaying and neutering our cats and by supporting societies that catch and release feral populations for purposes of spaying and neutering them and then returning them to their habitats. If a cat is well fed, then he will not eat his prey, which to me seems a shame, but in a way it is also preferable. Many modern cats have not had experience with mice as kittens and so while their instinct will tell them to catch and kill, they may not wish to eat it. If the kitten is from a barnyard society or other outdoor group, then it is very likely he had been fed freshly killed mice, and then almost dead mice, and then fully fierce catch-him-yourself mice by his mother. He will be used to the idea of a mouse for dinner and lick his lips when he sees one.

We had a cat at one time named Molly that would put her mice in my husband's golf shoes. It may have been a gift, or a joke, or she may have just been showing my husband her prowess as a hunter. A shoe is preferred to a pillow on your bed where one of my patients liked to deposit his to his family's great dismay. Or the doorstep, where another of my clients found one every single morning during the spring, summer, and fall, left there casually as a gift or offering, as the cat headed into the house after a night of hunting and prowling. Another one of my patients

Chapter 8

was very skilled with young squirrels and liked to bring them indoors to play.

Not all cats have the instinct or the interest to hunt; perhaps they feel it is beneath them. We had a clinic cat many years ago named Charlotte. She was an elderly little white Persian with severe arthritis and clicked around like she was wearing high heels. One weekend, when I was in the clinic to check on a few things, I saw a little mouse—one of those cute ones with the big ears—climb right into her cage and start eating out of her food bowl! It was quite a thing to see and I wish I'd had a camera with me. She did have stenotic nares and may not have had the keenest sense of smell, but I have to think she was aware of him since she was a cat. Maybe he did this every evening when the cats were put to bed. I don't know. Or maybe they were friends? And what was a mouse doing in a feline practice anyway? It was a very brazen mouse—a cat-like mouse.

My parents' cat, Bert, a most intelligent and engaging Siamese, played gently with his mice friends. He lived in an older house in the country over-looking the ocean and there was always a mouse or two to be found on any given day, usually in the basement. It was just a matter of carrying them up the stairs and through the cat door. He never did them enough harm that they couldn't scurry away when he was done with them for the day—at least at first. Eventually after a few days with a new mouse friend his cupping and re-cupping them in his paw became a game of juggling in the air and then to swatting them across the room like a volleyball where they would lay lifeless and of no further use to Bert. Often this sequence was so quick that my parents were unable to intervene. He didn't eat a single mouse in his very full life, but I think he might have been too full of other things. Bert was especially motivated by treats, but not to do tricks as many cats will do for treats. Instead, Bert had taught my parents that when he hopped on his stool that it was time for treats. Exactly five treats—one less and he may insist with his plaintive meow on the entire sequence of five treats begin again. Bert was a very charming cat.

Our cat Gus, the blood donor, developed an interest in my daughter's pygmy hamsters after one of the hamsters' first escape outside his cage. I think Gus also likely did catch a mouse outside after his own first escape, and suddenly after a year of barely noticing or caring about those hamsters he developed an obsessive interest in them. I found one of those little hamsters in the basement one day. Gus was nowhere that I could see. I picked him up and put him back in his cage. One of the tunnels had broken open and so I clicked it back together and thought he must have made the long journey down two sets of stairs by himself. Silly me! The next time, I found him dangling from Gus's mouth. Gus was very angry with me, but did let him loose unharmed when I insisted. Their habitat was moved to the spare room and the door was kept shut at all times. But eventually both the hamsters met Gus again, and they did not survive. Cats and pocket pets don't go well together. We still hope that those hamsters have found a new niche in the skeletal pathways of our home, but it has been some years now. I think Gus did them in and I can hardly blame him. It was his instinct after all.

We had a cat when I was growing up that was a very avid hunter. He was a ferocious predator. We summered as kids in a seaside cottage on Prince Edward Island. We'd never had a cat before, just a black miniature poodle with a tendency to stray, until one summer when the cottage was overrun with mice. It was the sort of cottage that was in a state of perpetual building. In the 35 years that I went there it was never fully completed. The ceiling was made of rafters and the fake paneled walls ended where the ceiling should have been. After we'd gone to bed, each evening we could make out the scurrying outlines of mice running the lengths of the rafters and all around the periphery of the house. Occasionally a traffic jam would occur and one would dangle on the electrical wires that hung down crossing the rafters. This was the first summer that we kids fought about who would sleep on the lower bunks instead of the upper bunks. No one wanted an upper bunk anymore, but it wasn't until my mother found a mouse dropping in the bottom of her teacup that it was decided we needed a cat. A kitten from our uncle's barn cat society selected us. My father wasn't so keen on the idea of a cat and clearly thought this little kitten wasn't up to the task. I secretly agreed and worried that as second youngest I would be made to sleep on the top bunk all summer long. There is a hierarchy among children that parents are not always abreast of. It is not dissimilar to what is found in cats. As the second youngest, I had no rights and no skill to negotiate my rights even if it could be established that I had rights. As my skill for negotiation developed, so too did my siblings and so the top bunk was a certainty for me. My younger brother was removed from consideration by his mere sweetness and vulnerability.

Leroy was a very cute little kitten; he was shiny black, very confident, and the top bunk felt worth it. A cat is a wonderful concession for many things in life. We named him Leroy Brown (after Leroy in the song "Bad Bad Leroy Brown"). He was a strutter and was only ever ferocious with mice and other prey. It took wee little Leroy about three days to figure out how to catch a mouse and another three days how to efficiently kill one. Having come from a barn, he'd likely had some practice already. I was disgusted and proud of him all at once. He made fast work of the mice in that cottage and then he took his skills outdoors. As he got older he also downed birds, bats, squirrels, rabbits, and probably other things that I didn't know about. Little wonder that he loved going to the cottage. Leroy, whom my father just called Puss, became my father's cat. My father was very determined not to like him and so Leroy spent most of his time charming and adoring him. Leroy lived a very long, busy life full of mischief. I tried to take him to university with me, but he missed my dad too much and I had to take him home again.

Cruelty and Kindness

I was not prepared for the amount of outright cruelty towards animals that I would witness and or hear about as a veterinarian. A kitten in a microwave, another in a clothes dryer, a cat chemically burned, a cat drugged by illegal substances, a head or a limb crushed by a door or a foot, a kitten thrown from an apartment balcony, a bullet, an arrow, a shovel, a rope, sticks, and stones. Thankfully, the brain does not allow retention of these images so well and what is left is just the impression that one must always be vigilant for signs of it because it is not always so obvious and there is not always a witness to it. Cats are innocent, and unfortunately, violence does not discriminate amongst its victims.

I remember one client that had begun to concern me as the kitten approached his final set of vaccines. The kittens name had been Butthead or something of that nature, which always throws warning signs up for me, and he had made some off comments that alarmed me. He was reluctant to pay for deworming medications for the parasite eggs we had detected in the kittens' fecal floatation test and he said that if the kitten had been smaller when he had found out about this extra charge, he would have flushed him down the toilet or maybe thrown him off his balcony. I somehow felt at that moment that perhaps he had done that kind of thing before and that he just may do it again. Luckily, the owner surrendered the kitten willingly to us when asked and we found him a wonderful home.

I was equally unprepared for the generosity and caring of so many people towards the creatures we share this world with who cannot look after themselves. These images are much easier to call up; the many clients that take on the stray even though they already have a few cats of their own; the client that brings cat after cat in for spaying and neutering from a stray population; the many clients that feed and medicate starving and sick felines that have strayed to their doorsteps; the many clients that bring their poor relations, friends, or neighbours in with a sick pet needing care and then pay for the visit and medications; and the group of college students who pay for the treatments required to save the greenhouse cat and continued to pay for special foods and medications needed to keep him well. I believe these same individuals spread kindness and light in all walks of their lives—kindness being their religion.

The cats themselves added such beauty to my workdays. We had a nursing mother cat that lived at our clinic at one time. We eventually found homes for her and all her kittens but while she was with us, someone came in with a two-week-old kitten that had lost its mother. She'd had just the one kitten and had been hit by a car and killed. Our mother cat didn't even blink an eye when we put this younger kitten at her mammaries to nurse. She adopted that kitten several weeks younger than her own and we eventually found a home for her too.

Another patient we knew had awakened her 12-year-old owner when the house was on fire, which allowed the entire family to safely escape. Another had prodded her owner's chest and smelt her breath repeatedly until she went to see her doctor and it was determined she had early lung cancer. The owner had recalled reading something about dogs sometimes being able to detect illness and decided to follow up on it. And, of course, Gus, who was my blood donor cat. These clients and cats are my heroes.

Chapter 8

Aggression in Cats

Aggression is a huge topic in cats and it is also something that does not discriminate amongst cats. The mildest sweetest cat may show aggression under certain circumstances. Sometimes aggression is a normal response. It is a part of the cat's language that is rarely expressed, but when it is, it is not subtle. The angry cat can be very dangerous. If you find yourself faced with an aggressive cat ready to attack, get something between you (like a door or a blanket) as quickly as you can. Do not try to handle or negotiate with an angry feline. An aggressive cat in the home can be a danger to you and anyone that visits your home. Serious injury and infection requiring medical attention can result from their bite or scratch. Do not wait to consult with your veterinarian about aggression when it happens, no matter how it presents itself.

Aggression comes in many different forms. Once again, the cause is generally multifactorial and will always contain a genetic predisposition. Rearing and early experiences also play a key role. Overall, rearing, nutrition, general health, and environment play roles in cat aggression. Sometimes aggression occurs due to injury or health issues that pique irritability or cause pain or discomfort in the feline. Sore teeth, arthritis, a bladder stone, urinary tract infection or inflammation, constipation, painful matting of the coat, fleas, itchy skin, anal gland impaction, or an earache may cause aggression. An overactive thyroid gland, seizure activity, and diabetes are other possible causes. Your veterinarian will need to rule these causes out before a behavioural cause is decided upon. Sometimes when there are mild aggression tendencies in an individual cat; the signs and frequency of aggressive events will increase when there are health issues. Unneutered males and nervous poorly reared or socialized cats will have a greater tendency towards aggression.

The most common form of aggression seen in a veterinary clinic, as you might guess, is fear aggression. Cats are control freaks and occasionally a cat coming to a place like a veterinary clinic will immediately sense total loss of control and so will react. This same cat may be a complete mushball at home, but will be a ferocious tiger in the vet's office. By degrees, they can learn that it will be okay, but it takes some effort. The best way to prevent it is frequent trips in the car that end other places and are accompanied by pleasant things like lots of love and attention, treats, or play time afterwards. A few trips to the vet that do not end in an actual exam or needle can go a long way as well to establishing trust and eliminating fear. Some cats do need a little sedation, anxiolytic medication, supplement, or pheromone spray or wiped on the carrier walls to help them feel good about it.

In the clinic setting, signs of aggression were usually half-hearted. I knew they often didn't really mean it and most did not feel so out of control. But I was being put on notice, which is fair and appropriate. I remember one big fluffy fierce cat that made a terrible scene every time he came, but once on the exam table he became so blasé. While I tried to find his body within his huge coat, he would slowly turn his head already bored with the scene and casually almost lazily take my wrist in his mouth and just press down hard enough so I knew who was really in charge. Fear in the feline patient usually presents as a catatonic cat lying curled on the exam table with his head tightly tucked into his armpit or chest. If he can't see me, I can't see him sort of reasoning. This is why placing a blanket over the head of a frightened cat can calm them significantly for examination. Many felines visiting veterinary clinics are there because they are injured or ill and so it is expected that some will react aggressively. Fear releases endorphins, which can actually lessen pain and so it can be surprising sometimes for the family to see their sick or injured cat act aggressively. At the same time pain and irritability from illness can lower the threshold for aggressive events so there is a fine line to manage sometimes with cat patients.

Cats have a very good memory for some things and so if they have had a negative experience with tall men wearing baseball hats, for example, then they may lash out in fear aggression at all tall men wearing baseball hats. Brooms or other items that resemble something they have encountered in a negative way in the past may on sight cause your cat to lose it.

Petting aggression is another common form of aggression seen in cats. It generally occurs when someone is petting the cat and suddenly, and sometimes quite ferociously, he will attack as though he questions their sincerity. We don't know exactly why this happens, but the length of time to attack is usually close to the same each time and it is thought that it is related to the period of time his mother spent grooming him as a kitten.

Chapter 8

Usually he does make a warning, a flick of the tail or his ears or whiskers might go back or he may arch his back slightly. And this warning is your cue to stop. If you can manage to stop petting him before the sequence of attack begins and give him a treat as a reward for not attacking, sometimes you can lengthen the period of time he will allow your affections before he attacks. One of our clinic cats, Coyote, has petting aggression and the time before attacking is pretty brief, except with men. She will allow men to pet her all day long, but we would have to tell our female clients not to pet her. She would only swat, not do any real damage, but it did seem rather unwelcoming in a welcoming cat. Her favourite place in the clinic was sitting on top of the reception desk. There may have been a bit of fear aggression in there too, perhaps as a result of her previous life. Coyote was chased into a client's barn by a coyote and that is how she got her name. They couldn't keep her because of allergies in the home. She also has a long grey coat like Beau that she will not allow to be groomed, and so we gave her a lion's cut (which became known in our clinic as a coyote cut) a few times a year. What she loved most was to sit on the reception desk and watch clients come and go. She was our greeter, but she was the china doll kind since she mostly only allowed clients to look but not touch. She's a funny little cat.

On the other hand, play aggression is when a kitten and then the adult cat gets over stimulated when playing and play becomes very aggressive. It may be fun at first, but then the kitten or the young cat latches on to your arm and starts kicking with his back feet, claws fully extended and digging in and he is biting and also using his front claws that dig into your flesh and stick there like needles. You peel them off one by one, but he is still kicking and biting and won't let go. Or maybe he starts launching himself at your legs and runs up you like a tree biting and scratching as he goes, with a maniacal crazed look in his eyes. This occurs in cats that have not had adequate time with its mother and littermates. A kitten learns appropriate play from its early playmates and mother cat. When he bites too hard, no one will play anymore, or his mother may swat or spit at him if he is unruly, and so he learns appropriate play. It is difficult for cats to learn certain things after those channels of learning are closed, but with patience and lots of positive reinforcement we can improve on things. With these cats there should be no rough play ever. You should play with him for a period of time just short of him becoming aggressive and make sure to give him a treat for appropriate play. Distraction can be helpful when he becomes aggressive. This is when learned tricks can be helpful. A short time out may be necessary to cool him off at times.

Territorial aggression is a huge problem in indoor cats and it can look like a couple of different things. Aggression between a cat and a human or other cat housemates, or whatever the cause, should be addressed as quickly as possible. Visit your vet right away to get it sorted it out. Sometimes territorial aggression becomes redirected aggression; for example, the cat is angry that another cat has entered its realm outdoors and because he is not able to attack the cat since he is confined inside, he will attack a human inside instead. This may happen right away or as soon as the human enters the room. The cat will remain sufficiently agitated by another cat outdoors for a certain amount of time. Once this happens a couple of times, the response by the human, which will likely be one of fear and flight, reinforces the behaviour and then the cat will often attack this human every time he sees him. It all began as redirected aggression, but then became directed aggression. The saddest case of this I ever saw was a young boy who dearly loved his cat, had been close to his cat, but then became a victim of redirected aggression. Before the owners sought help, this boy was being attacked a few times a day by his cat, every time he came into a room where the cat was or vice versa. In situations like this where there is danger for a human, you cannot be too careful. Your veterinarian can help you by various means, but safety for the humans will be the number one priority.

Redirected aggression can also be directed toward another cat in the house. Inter-cat aggression in multi-cat households is usually dominance aggression but can be redirected territorial aggression. This is also a difficult situation because the victim will be severely traumatized after an attack from a cat that had previously been a close friend. He will act like a victim, which in turn causes more attacks. This sounds like bullying in the schoolyard, doesn't it? With patience, a few behavioural modification exercises, and usually drugs, this too can usually be turned around, but see your vet right away about it—don't wait. When aggression occurs suddenly after no previous history, and if we cannot discover an obvious cause either medically or in the home environment, then I always suspect redirected aggression. Fluffy was a little Himalayan cat that suddenly would not allow our client's husband into their bedroom. It is hard to say what happened there, but in any case, the clients split up shortly afterward.

The one thing that is always true when your cat acts upset or aggressive, no matter what the cause, you will only reinforce the behaviour if you try to settle him down by cuddles, treats, and attention. He will think you are congratulating him on his proper etiquette. The best thing that you can do is calmly walk away. Let him cool off, don't engage with him, or try to get him into a room on his own for a short time out if you can. Yelling and throwing things at him will just agitate him further and may cause his behaviour to escalate. For example, the little boy would always give his cat some treats some time after the attacks once he had settled down. This very likely reinforced the behaviour. This same principle is true for all forms of aggression including fear aggression in the vet clinic. We want to reassure them that everything will be okay but in doing that we just make it worse for the next time.

Treatment for aggression will always be multipronged. In addition to behavioural modification exercises and environmental enrichment strategies, your vet may prescribe medications to enhance serotonin levels in the brain. It appears to me that aggression in humans is not so different. It also takes many forms. The root etiologies or causes are likely very similar but of course its manifestations and potential for destruction differ in scale.

"All cats are grey in the dark," so they say. Even a cat has a dark side of the street—those dirty little secrets that even we humans don't like to discuss, like hairballs, a fondness for junkfood, parasites, aggression, and eating mice or other prey. Yes, all cats are the same in some ways—all are grey in the dark—and can suffer from the same ailments, but it is our jobs as their owners to shine a light on the dark side in order to protect them from these dark things.

Watson

Chapter 9:
Grinning Like a Cheshire Cat

"Grinning like a Cheshire cat," means to grin broadly, or in other words, give a wide smile. The best-known use of the phrase is in Lewis Carroll's Alice's Adventures in Wonderland, published in 1865. There is no convincing explanation of why Cheshire cats were imagined to grin. Lewis did not coin the phrase himself, because there are several other writers before his time that wrote about "a Cheshire grin." It is more than likely that Lewis heard that Cheshire cats grinned and adapted it to his story.

Alice says to the Duchess, "I didn't know that Cheshire cats always grinned; in fact, I didn't know that cats could grin." "They all can," said the Duchess; "and most of 'em do."

Thinking about the magic I experienced with the cats in my practice over the years makes me grin like a Cheshire cat. I wanted to share some of the magic I experienced with my feline patients and some cat tales, (pun intended), and put a grin on your face too.

I think there is a little bit of magic in a cat. It would make sense that they would be five or six steps ahead of us in their thinking patterns since they live their lives that much faster. Okay, well perhaps their thought processes are not so advanced as ours; they go on feel more than thought but they still get ahead. Humans perhaps use too much time in thought processes that no longer serve them and get stuck there. A cat starts fresh every day. I frequently touched my forehead to their foreheads in my exam room and began what I thought would be a brief cat-to-human discourse on how their day was going and how shocked they must have been to have their plans disrupted by having to come see me. Imagine the annoyance of being pulled out of a nap into a box, heaved into a car, and dumped onto my exam room table.

With foreheads joined, the cat and I would essentially space out. I would come out of it with a shake of my head not knowing exactly what had been communicated or how long I had been missing. The cat and I would look at the owner and be a bit embarrassed—like we'd been found out. My older clients grew to expect this from me but new clients would still have to get used to this. I never really got used to it—the idea of losing time like that. It was like the cat's energy field grabbed and held me there. It always felt like a kind of magic to me. I loved those moments. They left me feeling calm and energized at the same time. I've almost replicated them with meditation but not quite.

The entire clinic was like a time machine, or perhaps more like a space ship sailing on cat time. It was a magic place and I began to think anything could happen there. Appointments sailed by. Days sailed by. Records piled up and more time sailed applying myself to them. I got lost in those records just as easily as I did with my forehead on a cat. I would look up and realize everyone was gone. "What time is it?" "Oh dear, must get home." "Where were you?" "Lost, in time again, on my cat ship."

My kids dreaded the words, "I have to go to the clinic just for a few minutes." They were painfully aware of the time-dimension disparity. They knew a few minutes of my cat time translated into a few hours their time. They waited at home in their own slow boring time dimension. One at a time, the kids would occasionally spend a day or part of a day at the clinic with me when they were granted a mental health get-out-of-school pass from me. They claimed a second thing happened within the clinic walls. Apparently my voice changed. And the words that I chose changed. It freaked them out a little bit. It was that clinical side that they did not see at home.

There is another very lovely thing about a veterinary clinic. It is one of the only places other than our homes where pet owners are allowed to brag about their pets, talk baby talk to them, and immerse themselves in their love for them. I had one client who went to a great deal of trouble to dress her cat up before her appointments. She had quite an extensive wardrobe. You can't help but smile when you see a distinguished seasoned professor of literature or history with his little half glasses perched just so on his nose come in with his cat's full-size suitcase for a weekend stay, the cat in his wife's arms, and each of them take their turn saying good-bye and behave yourself and what not and tenderly, reluctantly handing the cat over into our care. It's even funnier when a biker or trucker, with those rougher facades, comes in and does the same. And then, whereas on leaving the cat at the clinic they are sad, they are so eager upon return,

Chapter 9

thinking they may have heard his meow as they wait in the reception area and then everything else blurs for them you can see as they take their cat into their arms and start listing all manner of endearments and asking strings of nonsensical questions that don't have answers: "Did you visit the queen?" "Did you make any friends?" "Did you remember to brush your teeth?" "Are you the smartest and most talented cat in all the world, even in Africa?" "Are you daddy's best boy?" And if the cat has been left for surgery, then there may even be tears or the glint of tears as they are left and picked up again.

I want to explain two things here. One is that I never ever spoke baby talk to a cat. I always spoke to them as I would a human whom I respected. This was just as funny to clients as some would find the baby talk. The second is that I refer sometimes to clients as pet owners in this book, as you will have noticed by now. It is an old fashioned expression that I don't even believe in, but forgive me the use of it from time to time.

My clinic was certainly a very special place; I sometimes thought it was a magic place, and that magic rubbed off from the many cat visitors. There is a staff photo that I have from many years ago. In fact, it is the last thing that I gathered up among my things as I was leaving my practice on that last day. I had to admit as I studied each of my work family's faces in that photograph that there was the look of a cat in each of us. Something about the eyes—an intensity or a focus—or perhaps it was just the fact that we were all together at our workplace. It was the first time I had recognized what my husband saw and I have to say I was rather surprised to find that he might be right. I am perhaps a cat woman.

There is a second very strange thing about this photo that was mysterious to us all. Everyone one to the right of the photo disappeared from our lives, one at a time, from the outside inwards: a few moved onto other jobs, one went back to school, and one moved. I was in the middle, the next that would go, leaving my practice, but at the time of our discovery of the mysteriousness of the photo, I did not have any inkling that I would be next. How could I be, since I was the owner of the clinic? It seemed obvious to me that I would always work at my practice. Instead, I assumed if any change was to occur, the people in the photo would start disappearing from the left side, leaving me in the middle or in a random way negating the idea of a pattern, but I thought I would most certainly be left. New staff members arrived to replace those that left. New photos were taken, but I kept this one tucked into a tiny clipboard hanging from the wall in my office. One of the girls had folded it back to show only the staff that remained, so it was creased in several spots with me now on the far right. In the end, I was the next to leave.

The clinic felt like a second home and everyone that worked there was devoted to the care of cats from the start of the day to the end. I especially loved my treatment room, which had an open design so that I could see all my patients all the time. And I loved my equipment and my tools, which were an extension of myself. As with any place, there were things I thought about changing. For instance, the little fridges in the exam rooms were on the floor under the counter and so cats would try to get into them every time you opened it to retrieve a vaccine dose. On the other hand, every time you opened it a kitten would also step right inside which makes them an ingenious capturing device. Kittens can be so slippery in an exam room. It is not uncommon to have them almost bouncing off the walls playing and avoiding being confined to a table, a pair of hands, and a stethoscope. They can be difficult to catch once you've lost them on the floor. Some very dexterous cats were even able to open the fridges on their own recognizing that it may hold some food. High shelves were also perhaps a design flaw because some cats are reluctant to be examined, so a cat will often opt to observe from above if he can. The shelves did make me feel a little silly. I'm certain I could see them grinning up there like a mischievous Cheshire cat knowing I would have some difficulty retrieving them.

Even perfect design cannot foil some cats in a veterinary clinic. Beau, our clinic cat, went missing for an entire day. Our clinic cats had free range of the clinic during the day. We thought probably he was just sleeping in his bed above his habitat where he spent a great deal of time watching us from above. But, at the end of the day, we couldn't find him anywhere. We searched the entire clinic high and low, and we checked outside in case he'd slipped out with one of our clients. We called him again and again until our throats were sore. We shook treat bags. We made promises. I even sent someone to check the Humane Society. He was just gone as though he had performed a disappearing act. We sheepishly sent out flyers to surroundings clinics and the emergency clinic. How does one lose a clinic cat? I sat down in my office chair, and I was tired, hungry, and late for bed. Then I heard a cat yawn from inside my office closet where I had already searched three times, of course. In the closet was a locked filing cabinet and inside the locked filing cabinet in the back of the bottom drawer was Beau, who was very unconcerned and fully rested.

There is often a very strong bond between human siblings and their parents; it is so strong that if you think of them, chances are you will hear from them in the next few minutes, hours, or days. They were likely thinking of you at the same time. Answering the phone to hear my sister's voice several provinces away, I can't think how many times I tell her that I was just thinking of her, or that I was just about to call her. There is a telepathic connection there that cannot be explained easily but has to be recognized as fact. My staff and I recognized this same phenomenon with our clients and especially our patients. One of us would comment on a patient, "Oh I wonder how he is doing these days?" And within a day or two we would see the client in buying food, receive a phone call relating to this cat, or see the cat for an appointment. When I mused on how one of my more challenging (aggressive) patients might be doing at any time, the girls would chastise me for thinking about him, not wanting to hurry him in for any reason. It was taboo to think of these cats or to utter their names. My technician often only minutes later would come into my office and in her deadpan way tell me so and so is on the phone followed by, "I'm not kidding," raised eyebrows and an I told you so expression on her face. A slow smile, a Cheshire smile, would form on her face when she could report no harm had come to the cat and the phone call only confirmed once again that a magic we had long ago noticed in our clinic. I think when there is a group of people who work so closely together with so concentrated a focus, the energy that swirls in that setting can set up this kind of magic or whatever you want to call it. You've also heard about things happening in threes and this too seemed to occur in our clinic. A string of three blocked cats, a string of three euthanasias, a string of three missing cats, a string of three abscesses, and a string of three cats peeing out of the box. Was this more swirling energy? I think so.

I spoke earlier of a brokenness that I sensed in many of my clients and how I began to feel it was a universal brokenness. It is a very special intimacy that occurs often between a veterinarian and her clients after years of sharing in the wellness of a beloved cat. It is a special relationship that I felt so honoured to share. A veterinarian becomes a confidante often, just like a cat. And it was through these moments of sharing that I learned just how many of us are broken or damaged in various ways. Perhaps we have been pushed down by multiple disappointments or trauma, poor health, or tragedy that has formed our sense of self and the world around us. Perhaps we are just lonely. Perhaps there is financial stress and we are working hard to make ends meet and never seem to get ahead. The clinic was a safe place for clients to talk about their day. This is the magic of the cat again. Defenses down, once the exam is complete, words and emotion often just tumble out. It helps to know this is a universal condition. We share this life; we all have wounds. We can help each other just by listening, like a cat. This is one of the many lessons I have learned from cats.

Camouflage: Cats are Masters of Illusion

Cats are masters of camouflage and illusion. Mine appear to be everywhere at once or nowhere at all. They enter a room with stealth, and they leave before they are noticed. They appear and disappear as though by some magic trick.

Part of the disguise is their hair. The best example is a tabby cat. There are four major tabby patterns: classic, mackerel, spotted, and ticked tabby, and then there are other patterns that don't quite fit within any of the major categories. What I love about a tabby coat is not just the bands, spots, stripes, and M on the forehead, but that each individual hair may also be patterned. Even a solid coat may have ghost tabby markings as well, which can be seen when the light hits the coat in a certain way, which is a link to their African Wildcat beginnings. They have amazing detail in those hairs. Imagine the genes required for such detail. The lines around the eyes, tails, whiskers, and even their slit pupils all break up the image and help them blend into the furnishings around them. I have often thought it a shame that humans have so little variety in their hair colour. Wouldn't it be awesome to be a tabby? I did at one point have highlights put in my hair, which I have found can look attractive on other people. However, mine looked like stripes, perhaps tabby-like, but because each hair was not ticked, smoked, shaded, or glittered, but just dyed in bunches, I thought the effect was disappointing. But I suppose we have less need for camouflage than cats and we can create illusion by other means.

The bigger part of the illusion is how cats move. They can gallop like thunder and meow their intentions when they want to, but when the mood takes them, they practice stealth, silence, and being unseen. This adds to their mystique and elusive nature. When my children were very small I copied the cat's ability to be everywhere at once or no where at all. With the help of relief veterinarians I was able to come into work a little late and leave a little early several times a week. On these days my children had no idea that I had been to work at all and they knew that my focus was entirely on them. I always thought of it as a cat trick. I was so thankful for the help that allowed me to be in two places at one time.

I met a cat once early in my career that I still think of as the master of camouflage and illusion. A Humane Society worker, who feared he had injured a paw, brought him in. He was just very slightly lame. I carefully examined this tough little cat and reported he was in perfect health with no sign of injury or pain. I do not remember if he was black, orange, or tabby, or long or short haired, but I do remember his only imperfection was that his right front leg had been amputated probably years before and he walked on three legs instead four, hence the appearance of a limp.

Although I do not think you would ever see a cat pulled out of a magician's hat, I do think you may see a cat pull any number of things out of a hat.

Cat Tails: Odd Couples

Cats have been known to make odd friendships. I had a client whose cat had a skunk for a friend and spent hours playing with him outside daily. I was never very happy about that when she talked of it, because in addition to the risk of being sprayed, skunks can carry nasty things like Rabies, but you cannot pick your cat's friends if they go outside. Someone also sent me a video clip of an owl and a pussycat playing together. The owl swooped down low, the cat leaped up high. They both came together in the boughs of a tree and rested there together, occasionally rubbing heads together. I would never have believed it had I not seen it. Our own outdoor cats seemed to enjoy sitting in the garden while our chickens pecked away in the grass close by, but when the neighbours much larger guinea fowl came by, they would stalk them and chase them away. Another video I received showed a pet dove cooing into the ear of a sleeping cat, and then tugging on the ear of the cat trying to wake him. Another client told me of a feral cat she had in her barn that she could not get close to, but that her dog had become close friends with. They would lie on the picnic table together in the sun in the afternoon. Another young feline I knew retrieved a hamster from his cage and carried him gently by the nape of the neck to his own bed where the two slept curled up together. My daughter also recently came home from her daily run and told me she saw a woman walking her four dogs ranging in size and shape and colour, all on leashes, and 10 feet behind them was a cat following them. She was angry she'd not had her phone with her to take a picture. Louis and Dan (my grey cat and black standard poodle, respectively) who once lived with my dear friend and client have a very special bond and understanding. They lay together on the floor from time to time and at least once a day one offers the other a kiss or a snuggle. Mostly they are apart, but they come together regularly, perhaps in sympathy, and remembering their lost loved one. Cats are social creatures and I have to think they get a lot out of those friendships just as they do with friendships with humans.

The Biggest Tail of All: Never Let a Black Cat Cross Your Path

Black cats get a bad rap. I've never met a black cat that I did not like. There must be something that goes along with the gene for black coat that makes them so amiable and the opposite of aloof. It is possible that they find more trouble than the average cat of a different coat. And perhaps this is why they are associated with bad luck. Black male cats go missing the same way white male cats have a higher frequency of heart disease, at least in my clinic. So, maybe there is something there in the genetics for black cats that creates the adventurer—the explorer.

Going missing isn't always about the cat, though it is frequently about the coat. Sometimes it is a different kind of selection that causes him to go missing. Around October 31, black cats disappear from our neighbourhoods. I know it is a difficult thing to accept, but I believe they are being taken for pagan rights and sacrifice or just plain cruel mischief— more likely the latter. I certainly began to recommend everyone keep their black cats indoors during the weeks surrounding Halloween.

I remember one little black cat in particular that made me wonder if they do have some kind of special powers. I was examining this sweet little kitten while the mother and her five- or six-year-old daughter watched from their chair. The mother cuddled the girl because she was nervous about what I was doing. I prepared the vaccine remote from the girl and mother as to not create more stress for the daughter and gave it without event. As I removed the needle from the kitten in one smooth motion that the kitten didn't notice, the girl screamed that I had hurt her. She held up her own finger from her chair several feet away, and there in the centre of her meaty index fingertip was what looked like a needle prick with a drop of blood collecting atop it. The girl's mother and I exchanged baffled concerned looks. Some things can't ever be explained. This incident may have been more about the girl than the cat, but either way, I've spent some time thinking about it over the years. I'm sure there was a reasonable explanation but none was offered. Perhaps the girl had pricked herself at home by some means, wondering what it would be like for her kitten. Or perhaps she had been snooping in the cupboards before I entered the exam room and found our needles and syringes or the sharps container. Perhaps that was the real reason her mother kept her tethered to the chair—to keep her from further mischief. In any case, we moved the sharps containers from the closed cupboards in the exam rooms to the treatment room where little girls could not find them. But I somehow think it had nothing to do with needles.

The Tail of Three Modern Cats: Merry, Watson, and Louis

I'd like to tell you about my three cats. Merry is a longhaired brown tabby cat with facial markings that make her look angry, but she is a ball of mush. She was somebody's unwanted Christmas present. The annoyed recipient of this gift dropped the tiny kitten off at the clinic just days before Christmas and I worried about leaving her alone for the greater part of the holiday, because she needed socializing. I brought her home where she has remained ever since. We tried to convince my brother to take her home with him after the holidays but Merry was shy of a few days in converting him into a cat lover. He is a very hard case. She is a skittish cat and mostly unseen even six years later, but she is a particular favorite of my second daughter and she is a welcome addition to our family. She is terrified of the dogs and is the most subordinate of our three cats.

Watson came from the last of many litters produced by a feral community fed by my technician before we were able to catch and spay or neuter each and every one of them. He spent five months at the clinic attempting to find a home. Finally, I took him home because we kept tripping over him at work. He is a very odd-looking cat (probably why he couldn't find a home), resembling the phantom of the opera, asymmetrically spotted grey and white almost like a Holstein cow. He is very social because we worked with all those from the feral litters to make them well adjusted. He is a wonderful addition to our family. Watson has a special interest in wool and finds mittens and socks from I don't know where, at any time of year, and carts them all over the house. He also loves pink fibreglass, which he finds and pulls out of areas in the basement and deposits all over the house. I can't imagine where he finds it because I've attempted to close off all areas of contact. Watson makes a big show out of not liking our dogs; he hisses and swats at them, and then runs away as though he is terrified. I know he is not; he is just being dramatic. Watson is a drama king. Merry is the cat that gets terrified for real.

Louis, a beautiful shorthaired grey tabby, belonged to one of my dear clients who passed away quite suddenly and left her six cats and two dogs

Chapter 9

in my care. We also adopted her two standard poodles as well as Louis. Since moving in with me, I have learned that he is a real lover. He steps into your lap every chance he gets, turns over on his back, and gazes up with pure sweet love for you. He would prefer to be in your arms with his paws wrapped around your neck or be carried around like a baby looking up adoringly at you than any other place in the world. There is a rather large, almost rat-sized catnip stuffed mouse that came from his first home that Louis adores and carries around with him a part of each day. When he has this mouse in his mouth, he makes the most peculiar meow; we call it a proud meow—his, "I have a mouse," meow. He is the only one of our cats that enjoys time with our dogs. But then again, he did live with two of them as a kitten in a very small home, so that may account for his comfort with dogs. Who wouldn't benefit from time with a cat like that? We've decided that he is a little dim, or maybe innocent is a nicer way to put it. He lumbers around slowly and with an odd gait because he has hip dysplasia.

As it turns out, Louis was the one of the six cats who opted to defecate out of the box for the many years he lived with my client. We had discussed the different methods to identify the culprit (isolation one at a time; crushed nontoxic crayons of different colours in the food of each cat can also identify the culprit) but she never did figure it out. I guess she didn't like the crayon idea. It was a shame because we knew this cat had colitis and would frequently have bouts of bloody loose stool for weeks at a time, which was likely the reason for choosing to go out of the box. This often happens when evacuating is painful or untimely. Now it is just a habit that can't be broken. It took only a short time to figure out that Louis defecated out of the box once he was in my home. Thankfully, he only defecates on cement in the basement and it is an easy clean up. Even as a veterinarian, I cannot manage to change this behaviour so late in the game. He often shows a desire to go outside and I think this could remedy the problem but he is a very slow moving cat and in addition he does not always seem to have his wits about him, and I would worry about him becoming somebody's lunch. His colitis has not returned since coming to my home. This may be because I have three cats, not six, and our home is a fair bit larger to accommodate them all. I suspect he was overcrowded in his previous small home with two large dogs and six cats. It would have been difficult to maintain clean litter boxes with so many cats, and this factor was probably a component at the root of the problem. All the litter boxes in his previous home were in the garage with access through a cat door and all were in the form of a large tote container with high sides that he actually had to jump into. This was hard for Louis. He found a solution that worked for him. He has a very low-sided litter box here specifically for him in my home, which he does use all the time for urination and sometimes for BMs, but he does favour the cement. He is a dear cat and it is hard to hold it against him.

I have discovered a few new things about Watson and all my cats since spending more time at home doing my writing. They spend the morning sitting in a sunbeam on the kitchen table, watching the birds at the bird feeders. Watson breaks away from the other two watching the birds and begins his mischief. I've discovered a new place where he finds the fibreglass, which is under the bathroom sink in my bedroom. He slips in the cupboard and out again with a pink chunk in his mouth. He flushes the toilet in the upstairs bathroom and watches the water swirl down. He rattles the doorknob, which is like a lever, to my daughter's bedroom every morning for 10 minutes or so. Eventually, some days he manages to open the door. He rolls pens and pencils up and down the upstairs hallway that he forages from around the house. He opens the drawer in the kitchen where I keep the catnip and cat treats by leaning over the edge of the counter and slipping his paw through the pull handle. He drinks water and tea from abandoned, stray glasses and teacups around the house, using his paw. He takes naps in clean laundry. He finds a sock for the day in the laundry. He jumps back onto the kitchen table and onto Louis's back, biting him very gently, but just hard enough to remind him who is boss. He then grooms Louis and then himself. I could set my watch by his schedule of activities. He does not do each activity each day but he definitely has a schedule that he follows. All three cats spend the entire afternoon napping in the living room, each in their own favourite chair, and always the same chair. During the evenings and weekends, when everyone is home, their cat activities are different, and mostly unseen by us.

My three cats, Merry, Watson, and Louis, seem to be a stable group of indoor cats for the most part. When considering their home environment,

Chapter 9

I realize my cats have lots of space and windows. They also have plenty of stimuli to keep them entertained. But what if their view from the window in the kitchen were blocked (I now understand this is an important resource for them)? Or what if I forbade my cats' entry to the kitchen and they were not able to sit on the table? Lack of access to the kitchen would definitely change their schedule of events. Even Watson, who makes his own fun, would suffer for that. He would be stressed. Or, what if a new neighbour moved in, with several outdoor cats that frequently came into my cats' line of vision? That would be a huge stress for them. That would precipitate a problem for each of my three cats. Watson would be the most upset, and then because his threshold for irritability would be reduced, he would possibly take it out on his housemates. He may exert his dominance or display signs of his dominance more often. He may demonstrate redirected aggression and lash out at them. Merry would hide more in response. She may begin to have symptoms of Feline Interstitial Cystitis again (she had a few bouts of this when Louis and his dogs first came to stay with us) and begin to urinate out of the box again. Louis would be attacked more often. His colitis would begin to resurface and his stools would be runny and often with blood. All of these things would occur due to their environment and largely due to the fact that they stay indoors only.

My cats are all indoor cats, but this past summer they all escaped through a loose screen several times until I found their escape route, and I fear next summer it may occur more often by some other means that they will discover. The longer you can keep your cat indoors, the less likely they are to go far, but with each outing they will venture a little further, and so of course there is some risk. I may need to accept this risk because I understand how enriching this experience is for them. I am thankful that they lose interest as soon as the day light hours become fewer and the temperature falls. But I do wonder if these escapes have as much to do with the numbers of individuals in the house as they do about the temperature outside. During the summer, the children are all home every day all day and in the cats' way. Their schedule of events is disrupted and it becomes like one long weekend for them to find alternative entertainment.

I love that cats are all so different. They share many qualities of course, but they each bring something different to a household and to a day at the clinic. They all make me smile. Whenever I start to take myself too seriously or to worry about things best not worried about, I think of a cat's head spinning and spinning as his eyes follow a fly buzzing haphazardly around a room, or I think of a cat carefully grooming his paws, face, and ears. He doesn't remember or care that I just gave him a needle in his bottom or that he thought very briefly about biting me but then thought better of it. He lives in the moment, he notices everything, and he enjoys everything. He is eternally optimistic and has joy in his heart. You can almost see the lips turned up like the mischievous Cheshire cat. He knows life is good.

Chapter 10:
When the Cat's Away the Mice will Play

"When the cats are away the mice will play." This poetic proverb describes when controlling persons or entities are absent and the subordinates then take advantage of the situation. We see this happen in every segment of society: at work, at home, at school, and at play. It is natural for humans and animals to behave this way. There is always evidence of this, too: a warm TV and an empty cookie jar, cat hair on the placemats, or lick marks in the butter that you forgot to put away.

I have often felt over the years that the cat has struggled to be a priority in pharmaceutical and other ancillary medical and laboratory equipment and supply types of businesses. This seems to be related to the cost-related chaos in veterinary products and services. However, once there is a full understanding of where cost comes from, then it should no longer appear chaotic. I was always uncomfortable about the fee side of things, so in order to accept the fees I charged, which were based on fee guide, I had to look at each of the components of cost within my practice and sort it out for myself. In turn, costs for a veterinary clinic filter down to their clients. There is still a lot of confusion for clients because they usually do not understand why their veterinary bills are sometimes so expensive.

The truth is, I relied upon these pharmaceutical, ancillary medical, and laboratory equipment and supply businesses to be able to do my job well, and they offered tremendous support to my patients and me. Any difficult conversations of payment or expenses were worth it because, in the end, I could not have done my job without them. And I would never want to.

When I first graduated as a veterinarian, the standard equipment found in the lab was a centrifuge and a microscope, and the diagnostic equipment was often limited to the otoscope, ophthalmoscope, x-ray machine, stethoscope, a basic tonometer (measures eye pressure) and maybe an EKG. Today, veterinary clinics routinely house sophisticated laboratory equipment that may include some or all of the following: blood chemistry machines; complete blood count, blood gas, and electrolyte analyzers; and other diagnostic tools including radiograph equipment (and now digital radiography is emerging), blood pressure monitors, and modern tonometers. More and more veterinary clinics also now have ultrasound machines, endoscopes, and dental x-ray units. Referral veterinary hospitals often have extremely costly MRI and or CAT scan units as well. In addition to diagnostic equipment, a veterinary clinic will also have an anesthetic machine, scavenger systems, surgical instruments, autoclave, anesthetic monitoring devices, dental equipment for cleaning and extraction of teeth, and sometimes laser surgical units. As with the diagnostic equipment, these devices also need regular maintenance, care, and someone knowledgeable to run them. There is also another family of equipment related to care of our patients that includes things like IV pumps and oxygen tents. Therapeutic laser units are also now becoming popular.

Our technology and equipment today is far more advanced than ever before. Along with all of this came answers and diagnoses within hours, which is such an advantage when dealing with a sick feline. However, in many locations, an outside laboratory often makes multiple sample pick ups per day, which helps to get answers quickly even when there is no in-house laboratory available. In most cases, we can wait that amount of time for answers and treat supportively in the meantime.

I dealt with a one-stop-shop sort of business for much of my diagnostic equipment and laboratory supplies. I even bought my computers and software from them, and they were amazing in that they provided so much support. At any time of the day, I could call and speak with an internal medicine specialist or pathologist about a specific difficult case. I could even send x-rays digitally to a radiologist on staff for consultation. It was like having a set of specialists in my back pocket. Times have changed the way we practice veterinary medicine, certainly within my 23 years of practice.

The only thing I loved more than my equipment was my pharmacy. I should probably whisper this, so I'd like you to read it that way! I love drugs. It is amazing what pharmaceutical companies have been able to do for felines in the past few decades. It's exciting. One of my favourite things was when one of our drug representatives visited us to present a new drug or nutraceutical just for cats. It used to be that nothing was for cats, but they are finally beginning to get the attention they deserve. The reps were always as excited as we were when they showed us a new drug's molecular structure, how it is absorbed, and how it works in the body and is metabolized, as well as the safety studies, dosing protocols, indications, contraindications, possible side effects, possible drug interactions, and precautions. Intense research, design, development, and testing goes into every single pharmaceutical. It is an exciting time for pharmaceuticals, but it all comes at great cost and so a veterinary pharmacy is another huge expense for a clinic.

During the odd spare moment between tasks, if I was not admiring and fondling my patients, then I was facing my pharmacy admiring and sometimes fondling my drugs. I can still picture that beautiful library of drugs now: bottles and bottles of little pills of various colours, shapes and sizes, liquids, gels, pastes, chewables, and transdermals. Several columns of shelves organized impeccably by my cat-like technician; my crash kit with all my emergency drugs; rows of antibiotics A-Z; nonsteroidal and steroidal anti-inflammatories, drugs for pain and fever, muscle relaxants, antihistamines, appetite stimulants, anxiolytics, behavioural drugs, sedatives, antiemetics, antidiarrheals, laxatives, deworming agents, flea products, drugs for asthma and heart disease, diuretics, various vitamins and supplements, nutriceuticals to support kidney disease, urinary tract issues, liver disease, anemia, anxiety or behavioural problems, calorie supplements, liquid diets for tube or syringe feeding; and my injectable drugs of all the various drug families; eye and ear medications; and various medicated shampoos and skin barrier topicals. I'm probably forgetting to mention a few. Some drugs were locked away and others were refrigerated.

The thing about drugs is that you need them when you need them and so they must be there ready for use. They are all special and are only taken off the shelf under very specific circumstances. They each do their own thing, and when they are combined with other carefully selected medications, they can work wonders. But you need to understand them fully, such as their dosing, their indications, how they are eliminated, possible side effects and contraindications, possible interactions with other medications. Only your vet can do this for you. For example, when presented with a small cat with fever and dehydration and anorexia, the vet gives them a little of this, a little of that, some fluids, and often within the hour, that cat would be sitting up grooming himself and wanting to eat. Sometimes. It is much more complicated of course, because you need to know what you are treating, but it is little wonder that I loved those drugs so much. I never took their power for granted. A vet knows to use drugs sparingly with calculated selection, appropriate dosing, and only when needed. It continued to feel like a gift each time I took a bottle off the shelf.

The cost of care in veterinary medicine is skyrocketing. But that has seemed to be the case since I graduated from veterinary school. I don't know what can be done to contain it. If I look at each piece of equipment or individual service that I need or even that I provide, I can usually look at it and say, "Yes, this cost makes sense and is essential and valuable." But too often, the price makes no sense at all in comparison to other things. For example, why should a cheap drug specially formulated for a cat cost so much more than the human counterpart even when it is a fraction of the strength? But then on the other hand, why does a feline neuter cost less than an alternate monthly visit to the hair salon for colour and a trim? I am devoted to my hairdresser. I think of her as an artist or perhaps a magician and she is well worth the cost. She has a very faithful following due to her skill with scissors. But I often think when I am leaving the salon that it is a good thing that cats only need to be neutered once and not every eight weeks or so, because this is a service that cat owners too often do not want to pay for regardless of its importance, the breadth of knowledge and skill required to perform it, and its low cost. It is difficult when you try to compare apples and oranges, but clearly cost does not always make sense.

I think that part of the increasing costs in veterinary medicine comes from our growing expectations from our professionals. In other words, pet owners don't like the rising costs but it is pressures from pet owners and society in general that create the rising costs. One comforting result of this is the reluctance many vets now feel to leave a sick cat overnight in the hospital without supervision. It is becoming more and more common to transfer these cats to an overnight facility or emergency clinic so they can have overnight supervision. 99.9% of the time with appropriate measures (which usually includes the vet visiting at least once in the night) the cat is perfectly safe and in no danger whatsoever, and in many cases he would be happier in the quiet environment than in a busy overnight facility. However, vets feel pressured to send him off because of that remote possibility of problems. The client obviously wants assurance that everything will be fine; the vet cannot give this assurance with absolute certainty if the cat stays at the clinic unsupervised, so that is why they transfer the felines to an overnight facility. And in any case, it is the appropriate measure and I remember clearly when it all began. It was the beginning of the new age of veterinary medicine where each year brought new do's and don'ts and new costs to clients, all related to providing better care for pets.

Cat owners research online and sometimes have an idea about what might be wrong with their cat, and they try a few things themselves before taking their pet to a vet; when this happens, we are a few days further along that road of illness. Because clients are more knowledgeable than ever before there is a lot of pressure to do things better and better all the time. It is not just for the sake of wanting to do them better, but also because we know our clients and patients deserve and want that. In addition to this, the science of diagnostics, technology, and the practice of medicine advance constantly, which requires us to perform better and better. However, it is expensive to keep improving our methods. And it is impossible to go backwards.

I often feel that many pet owners want the best care, but only want to pay for archaic bandage methods, or not at all. Sometimes it is okay to not know exactly what you are treating, and to treat according to symptoms from the vet's best guess given the information that can be gathered from the physical exam and the information about what is happening at home. Information about what the cat is doing at home is especially difficult to gather because we often just don't know. Cats tend to hide or at least be less social when they don't feel well, and often we don't even know when the hiding or other behaviours began. We often don't know if they are eating or not, drinking or not, or voiding normally or not. Often this bandage type approach of treating and then waiting to see what happens can come back to bite you and can set you back another several days from treating the true cause of illness.

Veterinarians always want to practice their best medicine, but they also understand that costs can be high and they will usually be very open to trying the second best plan first if necessary. However, when a cat has not been eating for a few days, or has been suffering from dehydration or vomiting, then this kind of approach usually isn't appropriate. We need to gather more information so that we can get the cat back on the road to wellness as quickly as possible. Cats are especially vulnerable to secondary problems associated with fatty liver disease when they are not eating well. For example, when cats refuse to eat, sometimes it is the fatty liver disease that becomes life threatening for them and it is different from the symptom that makes the owner bring them to the clinic in the first place.

A veterinarian basically needs to be a dozen or more specialists in one. They need to be your cat's internist, anesthesiologist, surgeon (soft tissue and orthopedic), eye doctor, dermatologist, dentist, pathologist, radiologist, neurologist, cardiologist, endocrinologist, nutritionist, psychiatrist, oncologist, pharmacologist, and sometimes groomer. For most vets, they need to be all this for multiple species. They need to be able to spin all this together in their workplace and be able to speak intelligently on all these specialties both with clients and other professionals.

It is a very humbling thing to be in a room full of veterinarians. For my personal career path, I know I got to be a veterinarian and stay in that position by extreme focus, hard work, and pure tenacity. The brainpower in a veterinary school is an awesome thing to experience. I sensed a buzz of genius around me every second. I knew my classmates and professors could do absolutely anything, but had little choice due to their strong desire to work with animals. It's too bad that a vet doesn't also have superpowers or a magic wand. It is difficult to be a dozen specialists all at once without cracking. A vet is also just human; in order to do his work his office must double as a fully equipped hospital. A stethoscope is not enough. And this, in a nutshell, is why veterinary care is expensive.

A growing trend towards referring cases to trained veterinary specialists is happening when an answer cannot be arrived at easily or specialized surgery is required. This is a good thing driven by good medical practice, but again, it adds cost. I cannot go to a party or any kind of gathering without someone telling me about how much money their money-grabbing vet charged them for some kind of treatment. It's tempting to tell

Chapter 10

people I'm a zoologist or a farmer (since vets are a bit of both) instead of a veterinarian. I like it best when they don't ask. I listen and nod, and agree with them that, yes it is expensive, but their vet is not money grubbing—those are the actual costs. Then, I ask why they do not have pet insurance, because this is the real issue.

I wish it was different, but I've heard enough stories to know that people are often upset by what they spend on their pets. Yet, I don't know a single vet whose business plan is to ream money out of clients. I don't know many people who don't truly love their pets, so I'm not sure if it really is about money after all. I suppose in any field of work there will be a rogue or two: the odd vet who does tests that don't need to be done, for example. There are likely a few who are also incompetent. But, I think it is more often about lack of communication. I'm usually not surprised by the money amount they speak of because I know what it costs to run a facility and what the fee guide suggests for various treatments or surgery, but I am usually surprised by the vehemence and apparent lack of respect. There are always choices; costs are discussed and nothing can be done without permission, so I think this is often the crux of it—buyer's remorse can happen once the pet is home and well again, or once he is gone, because sadly sometimes they do not survive. But these sorts of comments also make me sad because this maligned vet no doubt is just like me, drawn to the care of cats and other pets because of his good and tender heart—not for lack of one. We don't get into this for that. Perhaps some vets don't communicate their tender heart very well if they have a tendency to be introverted, like many of us are, and perhaps they have poorly communicated what will be gained by a set of tests. If you cannot understand cost, then it is impossible to value it. I would also have buyer's remorse if I did not understand what I am paying for.

But I know how to turn a party around. Once my cover has been blown, there is nothing I like better than to talk about cats and my job, because both are awesome. Sometimes I even show my latest wounds inflicted by a particularly frightened or ferocious cat. I am unable to say a single swear word convincingly, but I have no difficulty at all talking about anal glands or fistulas, tapeworms, Cuterebra, or botfly larva poking through holes in the skin. I can say penis or vulva without blinking. And what vet doesn't like to talk about their latest shocking, juicy, putrid smelling abscess that they lanced and drained? My husband says I like to wait until food is served to talk about that, but I don't do it on purpose! It just takes time to warm me up and then I get carried away. Veterinarians make terrible dinner guests. Even though it is never my intent, talking about my work, which is not at all glamorous and can be very difficult, often acts like an antidote for those disenchanted with veterinary care. After hearing a little bit about the daily challenges of the job, most people usually feel that it should cost more rather than less. There is definite value there. A few of us vets just don't communicate it very well sometimes. We have to get better at that. The first thing I think of when I hear these stories is this vet has burnout from caring for pets and the cruelest part of it is that it gives others the impression that he does not care at all.

There is another little known fact about veterinarians—another secret that I feel compelled to share, and I hope that once it is said out loud that it will become less true. This secret is not just about compassion fatigue, and depression seen often in veterinarians. For vet students and practitioners, there is always that aim to provide quality of life and to prevent suffering in patients. For example, euthanasia stops pain and suffering in the pet. It is a practice that veterinarians are accustomed to. For the pet, death is not always the worst option. This knowledge, among other things, makes a veterinarian twice as likely to commit suicide as other health care professionals and four times more likely than the general public. It may cause the vet to feel like suicide is a viable option on a particularly bad day after many bad days, weeks, or months when pressures feel unbearable and there is no way out.

Possible explanations offered by Bertram in his March 2010 paper include high achieving personalities and perfectionism; stress beginning early in training and continuing through practice, long work hours, and insufficient support; isolationism in solo practitioners; high expectations by clients; high psychological demands; access to drugs and the knowledge on how to use them (Bertram et al. "Veterinary Surgeons and Suicide: a structured review of possible influences on increased risk." Veterinary Record March 2010). In discussions with veterinarians leaving schools today, many have very high student loan debt and much more stress about finances in general, and these are often combined with lower than expected income.

It is important for anyone considering veterinary medicine to consider all the costs involved including the costs of dollars, time, extreme effort, and challenges in veterinary school. It is a most wonderful profession and I do not want to discourage anyone from pursuing it. But it is only in a perfect world that we would not need to factor in expected remuneration, cost of living, life style, and paying off debts. We start off managing the costs, but then life often takes over, and before long our debt load increases. We've come so far, and it's difficult to turn back; after all, it is what we want to do for the rest of our lives. The hard work does pay off, but not necessarily in monetary wealth. It will likely not compare with other health care professionals with similar numbers of years of study and commitment. But, it's a pretty special job. It is more than a job, it is a career. The cats made it worth it for me. Now that I'm essentially retired, I'm always in awe of vets I meet or if I hear of someone going into vet school. I'm always impressed with them, which is quite funny because I never really thought of myself that way.

As for the cost of care for our cats and all pets, consolidation of resources into mega veterinary facilities makes a lot of sense. It would also ease the isolationism that so many veterinarians experience. But I think we need to be careful what we wish for. It is so human to do things in a bigger and bigger way when the masses demand it. My fear is that animal care would suffer and it would become more about running the monster facility than about the care itself. This might result in big businesses with the cat being "away," so to speak.

With the cat away, I worry a different sort of veterinarian would start to appear from veterinary schools—one who is less prone to having multiple pets, one who fits perfectly into their clothes, one who does not need to wash his hands before visiting the rest room—a hands-off sort of vet. This new breed of vet would need only please the facility but not the pet or the client. Something in between the two extremes is what is needed: a banding together of several veterinarians would benefit both the vet and the pet. This might slow the rising costs of fees and would hopefully provide a better income for the vet at the same time, because that side of the equation also needs attention.

Small veterinary practices still seem to be the acceptable and normal way of things, and personally I'm not sure that makes a lot of sense anymore given how many resources it takes to equip and run a modern veterinary facility. However, in small communities, this process will continue out of necessity to make sure that these communities have care for their pets. Even with an improved pooling of resources and sharing of talents between facilities, veterinary care will still be expensive. This is just unavoidable if we want access to quality supplies, equipment, and services. Nothing of high quality is ever cheap.

The way of the future is pet insurance. We have insurance for every other thing under the sun including our computers. If we value taking out insurance on our most prized possessions and irreplaceable belongings, then why wouldn't we also want to have insurance for our dear pets? We are used to this idea in human medicine, and this is probably why veterinary medicine appears to be so expensive. It is because we don't have a good idea of what our own true human medical costs are. Veterinary supplies are even more expensive than human supplies because they are considered by manufacturers and suppliers to be specialized, which of course they are. However, it is just a supply and demand equation. Veterinary products (drugs and surgical, laboratory, and other supplies) have smaller markets, and so the laws of supply and demand require a higher price. We may never catch up.

Pet insurance does several things. It allows cats and other pets to get the treatment they need when they need it. It allows the veterinarian to use their skills to treat their patients instead of euthanizing them for humane reasons. It allows the veterinarian to finally be compensated in a fashion in line with other health care professionals. If we do not embrace it, pet owners will continue to pay high fees for care and vets will continue to feel at odds with their work.

There are already multiple insurance options available for pets at very reasonable monthly premiums. It is becoming more and more mainstream to own pet insurance. Obtaining insurance when your cat is still a kitten or healthy young adult makes the most sense. If health issues exist already then insurance policies cannot cover those issues. The first few years tend to be the most costly, and then as seniors, costs may increase again as cats succumb to normal age-related issues. However, I can think of many cases where insurance saved pets from certain death or at least saved their humans a great deal of money and stress that would have resulted from struggling to manage veterinary bills. Urinary obstruction; surgeries to remove foreign bodies; fracture repairs; and treatment for poisonings, pancreatitis, and pyelonephritis are a few examples that would be extremely costly if not covered by pet insurance; this also includes anything that may require diagnostics and several days of veterinary care. Of course, these kinds of occurrences seem to only happen when we can least afford it. The stress of having a sick pet is difficult enough for many, and it is very overwhelming to add to that the stress of paying for it. But even other health issues that are less of a life or death situation that many cats just live with (as humans do) can be costly over time, and insurance can make these more manageable and stress free as well. That is really what the insurance is for. It is a kindness to yourself and gives you the opportunity to bring your pet home again instead of making that euthanasia decision, which is far too commonly made when there are no funds to support treatment. I believe every pet should have insurance. The common sales pitch is that it costs about the same as a cup of coffee in a paper cup every day. At the time of writing, for just over a dollar a day, you can purchase a comprehensive accident and illness policy for your kitten or cat. Make your own coffee, skip the doughnut, and buy insurance for your beloved feline. I agree with this line of thinking, and many clients have shared their sense of relief when they listened and did exactly that.

What bothers me the most about the rising cost of veterinary medicine is that there is so little leeway in the case of a kind-hearted soul who brings in a sick or wounded stray cat for treatment. This good deed often turns into an expense. Cats settling and reproducing in barns is another situation some farmers find difficult to manage; spaying and neutering would solve the problem, but this quickly gets out of hand cost wise. We see the same kind of situation with owned cats when finances are tight. If a veterinarian offers his services at little or no cost every time he feels compelled to do so, then he would quickly lose his shirt, have to send his staff home, and close his doors.

Care for Cats is a program that helps bring awareness to the problem of overpopulation and notes that people, not cats, are the problem. Each feral cat has its own story of how he came to be there. If only they could tell us, but we understand enough only by their wariness. Similarly abandoned individuals band together for safety and company. Each new kitten born into a feral community is increasingly wary and wild due to neglect, inadequate population control, and the human disposable mentality—not of their own choosing. Feral cats may never make good pets, but we can still help them and we must. We can start helping them by taking care of them, and then by ensuring that their numbers do not continue to climb. We created their plight and we must also be the ones to ease it. With early intervention, many of these kittens and adults can learn to enjoy human contact and eventually make good pets. Early intervention means fewer litters being born. Care for Cats is a very thoughtful program geared towards increasing the value of cats everywhere by joining all sectors in a community: veterinarians, municipalities (animal services, public health), animal welfare (large associations and small rescue groups), and the retail/vet industry. Together we can send a unified message that cats need to be spayed and neutered, identified, registered (like dogs are), and receive routine preventive healthcare. See careforcats.ca

Certainly, veterinary medicine is expensive. Obtaining insurance right away will prevent a lot of stress and worry through times of illness. Your focus can be on your cat's recovery instead of the cost of recovery and what you will have to give up in order to treat your cat. Maintaining a good preventative health program that includes regular checkups, environment enrichment, spaying and neutering, good nutrition, and attention to details like weight management and oral hygiene will go a long way in preventing costly veterinary care.

Good Salesmen

My husband used to tease me by saying that a cat can sell anything. Once again, my husband was right, though I don't think I would have noticed until he said it that cats are everywhere. Every Christmas, birthday, or any other gift-giving occasion brings items relating to cats into my possession.

Chapter 10

I have cat tea towels, sculptures, lamps, clocks, candles, candleholders, music, videos, paintings and posters, ladles, napkins, napkin holders, paper towel holders, plaques, blankets, rugs, Christmas tree ornaments, sun catchers and other stained glass, wine glasses, mugs, plant holders, measuring cups, necklaces, watches, earrings, charms, underwear, nighties, t-shirts, sweaters, vests, socks, place mats, and board games. You name it—it comes in cat. Among all these cat items, I cannot find a single mouse!

A day does not go by without something cat related in my Facebook or email. People just send me stuff like their photos, videos they've found online, or funny sayings about cats. Videos of bathing cats, medicating cats, of cats beating up or chasing dogs, and of cats grooming kittens or dogs or bunnies or even birds. Recently, someone sent me a picture of a large crowd gathered outside someone's house trying to catch a glimpse of a famously obese cat that lived there. This encapsulates the power of a cat. There is a plethora of material out there on the subject of cats. Even greeting card companies solicit the use of cats in their products. The Broadway musical CATS has been a long-time success. My ringtone is a cat meowing.

Add to that the cat's popularity in commercials selling everything under the sun, film, and even the news. Cats are excellent actors and add softness, whimsicality, and legitimacy to the otherwise ugly business of advertising that only babies can rival. Have you ever seen a commercial with a cat that you did not like?

Over the years, I've heard my husband say, "See?" many times as another such commercial or movie uses a cat or kitten to produce an effect. Cats are very effective communicators; they are salesmen.

My son says that cats are a cultural wonder of our time because they serve no purpose for us whatsoever and yet we fall head over heels in love with them and cater to their every whim. He says a dog hunts and protects, along with many other important functions. I reminded him of the Great Plague and how the rats carrying the fleas that carried the plague took over because cat numbers were down. And he reminded me that there are also breeds of dogs that are especially good at ratting. Maybe he is a little bit right, which just goes to show what good self-promoters cats are. They have made themselves indispensable to us, yet we cannot exactly tell why.

Look how our felines, the ultimate salesmen, have charmed their way into our homes to be fed and spoiled by us. We are slaves to their bathroom habits and will do anything to keep them healthy, happy, and sitting pretty on velvet and lace.

I have never gone out into the world looking for a cat. Each of them have found me. Mind you, it is somewhat easier to do so in a veterinary practice where there is always a cat looking for a home. It was also a common response from my clients when I asked where they had found their cat: "He found me!" Well, of course he did. I remember one client of mine who fostered a cat for four or five years and eventually after some years of knowing this pair, I suggested that perhaps he really was his cat. Oh no, he said resolutely, he didn't have any of his own cats. The cat seemed to object to this idea as much as I did as he brushed up against him, purred, and rubbed his cheeks on his face. I told him, "The cat may not be yours, but you are his." This was a very special cat of which I was extremely fond. This cat toppled over the treat container every time he came to see me, and I never remembered that he did this until he had done it again. Perhaps his beauty distracted me from his mischief. This cat noticed everything: the pictures on the wall and those dangling from the ceiling, and the cats on the fluorescent light covers. He watched my every move intently, and he was always so interested in everything. He had contracted Feline Immunodeficiency Virus early in life and his foster family had taken him from a rescue shelter and given him a home. He thrived in their care.

Cats are good self-promoters just by being friendly and confident and going with the flow. They don't push too hard. It is soft sell at its best. They are the perfect hook. They are charming, eager, and attentive, followed by a period of aloofness, and then we are hooked. A cat is good at being in the right place at the right time and seeing opportunities when they arise. They know that every place and every time is the right time to make things happen. The image of a cat in any context, commercial, packaging, or other promotional material is a subliminal message for success. We see a cat; we think success and abundance because the cat has a talent for both. And then we think, I want that.

Chapter 10

T.G.I.F.

Fridays, commonly known as Thank God it's Friday (T.G.I.F.), were always our most challenging days at the clinic. This is a phenomenon seen routinely in veterinary practices. For the most part there is nothing sinister about it. It comes from hoping for the best most of the time. I used to call them my almost dead days and that is what they were. We did not even book routine appointments most Fridays because we knew there would be an onslaught of extremely sick cats looking at us with that T.G.I.F. look on their faces. And I understood in an instant, as I looked in those huge pleading eyes, what the past many days had been like for them. For the most part, Fridays were busy because as the weekend approached, families became concerned that waiting over the weekend wasn't a good idea for their pet, and that they had waited long enough for their kitty to feel better on his own. However, there is a second kind of client that one sees on a Friday that causes heartbreak and disenchantment with mankind. I feel I must tell you about it because otherwise you could not know.

I could never quite figure it out. I began to wonder sometimes that if waiting those extra few days somehow gave permission to not go ahead with treatments of any kind, and instead go straight to euthanasia. If the cat was just that much sicker or closer to death, then it was easier to make that decision without guilt. Perhaps some clients hoped he would die peacefully on his own and they would not need to intervene in any way. However, the cruelty of it settled in to me; the dehydrated vomiting kitty that had been vomiting already for over a week and now could barely stand, the blocked cat dazed by severe pain that should have been seen two days ago, or the cat that hadn't eaten more than a few treats for nearly two weeks and was severely jaundiced from secondary fatty liver disease. Fridays brought clients that were angrier about the vomit stains all over the house than they were concerned about the suffering of their pet, and they were also angry that it would cost money to make the vomiting stop.

This is how I spent my Fridays: I tried to put cats back together that would have had a much better chance had I seen them on the Monday earlier in the same week. In order to get through it, to keep from screaming and crying, I defaulted back to the fact that at least they had finally brought them in, and didn't prolong it for one more day. But isn't that only the very minimum—to finally bring them? What if we did everything that mattered in our lives like that? Would you wait until Friday if your children were ill? Pets suffer in the same way. It is not just a cat; it is a living being made of flesh and blood, a beating heart, and the same building blocks as humans—the same vulnerability to sickness and pain.

C.S. Lewis said, "Integrity is doing the right thing even when no one is watching." A cat or a dog will not tell on you if you leave them to suffer, nor are they likely to award you a purple heart and write newspaper articles about you for your efforts. The act of kindness, the good deed, is also the reward. When someone takes the time and effort to help a helpless stray creature in need and is not expecting any acknowledgment or return on investment, this shows true kindness and strength of character. The cat, as a family pet, should be able to depend on it.

These T.G.I.F. situations did not arise with my regular clients; they arose with the transient client who only sought help at the very end likely because of pressure of others, like from their small children, or by a final push by their own conscience or to hide the lack of one. But they were frequent enough that it became difficult to cope at times.

In response to my suggestions for care, I received T.G.I.F. responses such as, "Can you promise it will never happen again?" or "Can you promise that he will make a full recovery?" or "Can you promise he will live the rest of his life without this issue?" I would not be able to make such promises. I was not fixing a car or dishwasher under warranty; I was trying to help heal a complex creature of God. There are no spare parts. And so they would ask for euthanasia, sometimes even when I knew I could help with medication alone. I would want to extract (but could not) a promise from them that they would never ever own another pet. But I knew they would leave and probably immediately go out and get another one—a prettier one, one that matched their own hair better, and was new and unbroken. To add drama, their visit became the next vet story that circulated among friends and family, and very likely to the next vet they met out in society, about the cost of veterinary care, which ended by the cat dying anyway.

Chapter 10

Some T.G.I.F. clients would laugh when we explained to them what needed to be done to help their pet and what the costs would be. They would then say those predictable words that vets have heard time and time again, "But he's a cat, he didn't cost me a penny, and you want me to pay you that kind of money to fix him?" And it wouldn't matter how much—any amount is too much for some people. Something that is free should not have a cost apparently. I never really understood that. What does what a cat's cost have anything to do with anything? One man told me his cat's stay at our clinic would cost more than a night in a hotel, but I very much doubted that his cat would find healing in a hotel. It is a ridiculous comparison. Nevertheless, I found that money makes people say and do ridiculous things, especially when it comes to their cat.

I know a family cat is valued beyond words; one can't put a price on a beloved pet. It would have been worth giving up a Christmas present that I had thought that I couldn't live without, or even a long anticipated family vacation, as long as I was able to have my cat come home from the vet. My pets now are certainly worth more to me than what we spend at the liquor store or coffee shops in a month. And that is usually the kind of costs we are talking about. When I would come home feeling contaminated by these kinds of interactions, my husband used to say that these sort of people spread their joylessness everywhere. They take it to the car mechanics, the computer experts, their dentist, their chiropractor, their orthopedic surgeon, their children's teachers and coaches, and their workmates. Some people don't take responsibility and they don't act accountably. I know my husband is right, but a vehicle doesn't suffer. A cat matters because it can suffer.

Nevertheless, I did learn one thing about these situations in my many years: the ones that really need the help with care of their pets are the ones that do not ask. We watched carefully for these individuals and tried to help where we could, and I always felt that it was so greatly appreciated. All we can do is our best; this is what I told my clients who were struggling over decisions trying to balance expense and likely outcome. And sometimes when outcome is questionable and treatment is lengthy and expensive, the very best we can do is to relieve pain through euthanasia. In situations where the choice is rent and groceries or new winter boots for the kids, then euthanasia may be the best we can do to prevent further suffering, and hopefully next time our best will be something different. But this should only be a last resort when all other options have been exhausted.

It was difficult for me to help people who expected or demanded free help, who made me feel hostage, and for whom it was a matter of the price of a cat's life versus forfeiting valuable time, treatments and medications. They made me feel that because I loved animals and knew what to do to help them, that it was entirely my responsibility. It left a bad taste in my mouth as I watched them drive away in their Porsches with their cat already feeling better and looking out the back window—always on a Friday.

A cat is a living, breathing thing; he hurts as we do and he suffers as we do. He is stoic; he is undemanding in his illness and he waits for us to help. Suffering does not wait.

I notice that the cat, sick as he is, has compassion for us both—his arrogant owner and me. The cat says don't take offense, I do not. I accept him as he is; he is a work in progress. He says this as he climbs into his carrier and his eyes move from me to his owner and back again. He tells me that even this is an opportunity to grow compassion. His second look my way says tenderly to me that I am also a work-in-progress; it is not good enough to have compassion only for the cat in the room. And he is right. I don't miss his slow blink with slight squint at the end, which says thank you to me. That is all the thanks I need. Those are the moments I miss now that I'm no longer a practicing veterinarian—the unexpected gift from the cat on a bad day that reminds me why I do it at all.

The idea that "When the cat is away the mice will play," is a reflection of imbalance. I do feel that the field of veterinary medicine is imbalanced. There is a lot of frustration within the veterinary community about costs related to the care they provide. They share the belief that veterinary care is essential to families and communities and they understand it is costly. The cost of providing care can also be high in terms of the vet's mental health in today's society. I think balance is returning as more and more families are investing in their pet's health with good preventative veterinary care, with spay and neutering practices, and by purchasing insurance policies. There is also a growing understanding of the true costs of care and an acceptance and realization that it is very good value given the importance of the cat in our families. Increasing focus on feline population control through programs like Care for Cats (Cat Health Canada) and catch/neuter/spay and release programs provided by many feral cat associations is also important to reduce the numbers of stray and feral cats suffering from and spreading disease in our communities. These programs need our monetary help to keep them doing their good works.

Watson

Chapter 11:
Conclusion: The Guru Cat and the Copycat

I think of a cat as a Guru—a teacher who can teach us how to live our best lives in harmony with our true selves from within, and also with all mankind and nature too. There is wisdom in the Guru cat. The yoga teacher does the same thing. It's funny to think of the cat as a Guru or a Yoga master but I do. I wonder if the first yogis in ancient times may just have been copycats. The origin of the word copycat appears to be from Sarah Ome Jewett's The Country of the Pointed Firs and refers to kittens imitating behaviours of their mother cat. A newer understanding of the word refers to humans who shamelessly use other people's behaviours or words as their own.

I believe being a bit more cat-like could lengthen and enrich our own more complex lives. Of course, the cat does not plan for the future or concern himself with progress because he is not human. You may argue that he may as well smell the roses. But the cat has his own cat size struggles. I don't want to say that I have modeled my life after a cat, because that would sound crazy, but in fact I have essentially done that. Certainly the cat was my first teacher followed by other wonderful written teachings of various spiritual gurus, and the life transforming lessons of Kundalini Yoga practices as taught through my own amazing yoga teacher. These teachings are so simple but also powerful, and they all bring me back to what I have already observed in the teachings of a cat: be kind to everyone. Never say a harsh word against yourself. Never say a harsh word against any other person. These are the yogi's lessons for prosperity. Rather cat-like I think. Start fresh every day. Assume the best. Don't hold grudges. It seems a lovely way to live. As simple as it sounds, it takes practice to follow. We humans are very good at disparaging ourselves and kindness does seem to be a muscle that we do not exercise enough, even towards ourselves. A cat inspires me to try to be my best me, and he also inspires me to surround myself with others who do the same.

For longevity (a long healthy life) the yogi simply says you are as young as your spine is flexible. Again, this sounds like a cat that loves to stretch, to move his body, and to stay fit in his natural setting. I think the only pose I've not seen a cat do is an inversion like a headstand, shoulder stand, or handstand—the yogi's prescription for many ailments. Well, maybe a tree pose and a few others would be tough too for the cat. But a cat is half way there to inversion anyway since he is on all fours and has his head below his heart often enough. We've all seen the cat do "downward facing dog" with his bum in the air and his front paws and head stretched out in front of him in a nice deep stretch of the back and limbs. That is an inversion cat style, even though it is named after the dog, which is a shame.

The cat has his own pose of course—the Halloween cat pose (where the cat's back is arched) is also copied from the cat to work the flexibility of our backs. There is pure science behind yoga, which is why it interests me so much. Every move stimulates something physiologically within us making us healthier. It is not just about flexibility and fitness. It stimulates our glandular, lymph, digestive, nervous, and circulatory systems as well. If you have thyroid issues, there are exercises for that. If you have digestive issues, there are exercises for that as well. A cat practices yoga and I believe it helps keep him well. As with cat nutrition I have found that not all yoga is created equally. Some teachers have a very strong understanding of anatomy and embody the entire culture of yoga and all its teachings, and some do not. It is not something I would ever do without a teacher, because they can guide you according to your needs so you do not become injured.

Some forms of yoga (like Kundalini Yoga) have specific chants that stimulate certain things that they want in their lives. Until very recently, this was another secret language of sorts reserved only to a few, but the science behind it is so powerful that it has now been made more public so that anyone can find this practice and benefit from it if they wish to. Chanting is much like prayer (or purring). The repetitive chanting of the beautiful words found in chants or mantras causes the tongue to hit meridian points within our soft and hard palates, which in turn stimulates areas of the brain to release serotonin. In addition, the vibration of the chant is healing, much like the purr of a cat. There is further intent here, and I would argue that the purr has the same intent. And that is the intent that these positive vibrations create a wave of pure healing and loving energy like a gentle tsunami washed invisibly over every soul. Each yogi and each cat in his place does this. This is one of the ways cats work their magic upon us.

Yogi Bhajan says, "Happiness comes to those who are happy because happiness loves to be where happiness is." This also sounds like a cat, which in his or her natural state is full of joy and wonder for life. He is always the same; his true self; like every day is his birthday. When I say a cat's natural state, I mean wherever he is able to fully express himself and fully be a cat. For your cat, this may be within the safety of your home. For many others, however, it means a space where his home range is larger than the size of a house, where he is able to be outside with his paws in the dirt.

Meditation is a part of the yogi life style as well. It is a way to deeply connect with your inner you and the greatness of the universe at the same time. It allows us to enter a space of stillness, and out of that comes a deep appreciation for life's blessings, beginning with the breath and life itself. When there is gratitude, then there can be true joy. When there is gratitude and we forward that feeling in all our interactions, then it can be contagious. Again, we can picture the cat in its meditative poses like Buddha, as he appreciates a shaft of warm sunlight. He is recharging himself so he can do good.

Following all yogi lessons, which are largely centered on love and compassion for all living beings and for mother earth, can leave you wide open to other people's moods and agendas; we are taught that compassion for oneself is also important. A cat can do that as well. A cat at times will practice tough love; he is always compassionate, but he will tell you when you've crossed a line. And sometimes he knows he must walk away.

I read somewhere very early on in my undergraduate degree, probably in some textbook, about the idea of life being measured in heartbeats; with each heartbeat we are literally counting down to a finite number of total heartbeats to the end of our earthly life. This idea has always intrigued me and I have looked for its proof throughout my whole adult life. It does make a lot of sense—smaller creatures like cats, birds, and mice all have much higher heart rates than we do and their time on earth is much shorter. When I met death in my patients, I couldn't help myself from wondering how many times their heart had beat, and what was the final count. That will be my own first question when I reach the pearly gates.

This theory falls apart when you think about species like the horse, which have lower than human heart rates and much shorter life spans. But I still feel there must be something to it. There must be other factors, like mass perhaps, or body temperature (a cat's is much higher than ours), or four-leggedness? Maybe the quantum physicists will work it out one day—the equation of life, that no doubt will include some constants and several variables, the heart beat being just one of them. It is one of them; I'm convinced of it.

Cats have a very rapid heartbeat compared to ours, and they have a much shorter lifespan. My stethoscope and I counted so many heartbeats day after day. It is still a marvel to me, that heartbeat—involuntary, automatic, tic after tic after tic. And it is precious. It felt like I was listening to my own heartbeat as I counted down each precious pulse until the end. It made me want to live better. It made me want to make each beat count. Those beats had their own special qualities; the youthful, eager beat; the tentative, shy beat; the agile, athletic beat; the confident beat heard through purrs; the tired, weak, and ailing arrhythmic beat; the angry beat heard through growls; and the nervous, frightened, and rapid beat that slowed to a normal beat as comfort set in. All these beats combined within me, each part of a larger collective beat like parts of my own self. It does not matter if it is human, feline, canine, or otherwise. Each of us carries a beat within us that unifies us. Our beats mesh perfectly together to form a continuum, just as our talents do. If we do not belong in one place, it is because we are dearly missing from another. We are all essential.

Wakefulness is another key variable to longevity and once again the cat has figured it out. He knows that hours awake is another clock ticking and so he takes his catnaps; with such a high heart rate, he needs to take many of them. I suffer from insomnia and bouts of it will last typically six to eight weeks no matter what I do, so this idea is particularly interesting to me. Sleep impacts our lifespan and certainly our quality of life. Lately I have been wondering if I am nocturnal by nature like my husband suggests, and not an insomniac at all. I have finally mastered the catnap. Meditation also is helpful for deep relaxation and is worth several hours of sleep.

I have already mentioned other feline traits that we would all do well to imitate in our lives for the sakes of our longevity, quality of life, and also for the lives of our loved ones. Staying active and keeping ourselves groomed are obvious ones. Other lessons are perhaps more subtle.

The cat purrs. He tells us to sing. The vibrations that singing creates within our vocal cords and our entire bodies are good for us and it is a joyful thing to do. Singing and listening to music can increase the feel-good neurotransmitters at the synapses in our brains. The cat doesn't need to know why; he just does it because he knows it feels good. The cat is present and appreciates every minute. He notices and appreciates the small pleasures in life. The cat says be present. Who knows when our last heartbeat will come?

Along with being present, the cat is curious about everything. He says be curious. Ask questions. Learn new things. But don't take yourself too seriously. The cat is confident. He says be confident. A cat is also very good at facing his fears. But, he doesn't always work it through. For instance, when he sprays on the back of your couch in response to the fear of losing his property to a roving cat outdoors, he didn't understand that that would upset you. He will never understand that.

For us, fear is a little bit different because it doesn't usually come and then go—it lingers, festers, and it worries us. Usually it is a worry or a problem that does not even belong to us. And that is what the cat is good at: knowing which problems belong to him and which do not. The spray on the couch isn't his problem; he's just marking his territory in the best way he can in response to a threat. It's his solution to a problem. He certainly couldn't imagine that doing something like that repeatedly might get him evicted. I don't know if a cat considers the worst-case scenario or not. I think my cats do when they send Louis to get past the dogs to ask for food in the morning. I like to imagine the very worst thing that might come from facing a fear or problem. Following these scenarios through helps you to realize that there isn't much we can't handle in life. Fear can be so crippling.

The cat is affectionate. He says be affectionate. When life gets busy, we lose the gift of touch. We lose touch with extended family and friends, and we even lose that closer touch with the loved ones within our own homes. We tend to put that focus on ourselves instead—what a difficult day we had, what a bad headache we have, how tired we are—and this inward focus causes our souls to shrink and shrivel. A lot of time can pass by before we realize we've been reclusive and antisocial. Think of the cat that jumps down from his chair (or your kitchen table) when he hears the car in the driveway. He's right there to greet you and tell you he missed you and he's glad you came home. I think the cat wants us to play it forward, but there is a limit. My husband says that he cannot take the place of thousands of cats, and he can't. Apparently, there is such a thing as being too affectionate. He'd like me to stop petting and kissing him quite so much; he says this kindly but he means it. The kids and the cats make the same complaint. And they are right; I miss my hands in those warm purring fur coats and those intense eyes with perfectly applied eyeliner looking upon me.

Cats forgive. They tell us to forgive. I benefitted daily from the cat's ability to forgive. They do not necessarily forget but they do forgive. After a few days of absence, your cat may give you the cold shoulder for a while but he does forgive you. In a veterinary setting I was always amazed how my prodding and poking and needle jabbing did not cause more grudges against me. It seemed I was forgiven almost immediately for such intrusions.

Cats are an addiction that I do not have to give up entirely. Writing, for example, is not so different than veterinary medicine. It is still about the puzzle. In medicine, you must put the pieces together to make your diagnosis and treatment plan. In writing, you start with the puzzle complete and try to create the same layers and intrigue for someone else to fathom through. In practice, at some point I noticed that my medical records were becoming an outlet for composition, but the vocabulary—though beautiful—limited full expression. Fellow vets expressed their fondness for brevity when required to review records to follow up on cases, so I returned to the S.O.A.P. standard of brief descriptions of Subjective, Objective, Assessment, and Plan, and I had to forget about the beauty of the creature and avoid descriptions of masses, for example, beyond exact location and size in centimetres. I have no such restrictions now. The cat is still part of this process of writing and is never far away as there is always a cat on my lap. But, I can spread some of my attentions onto my characters which seems to suit everyone nicely.

A cat tells us to not make assumptions or be too quick to judge. We cannot know what other people's lives are like or what their struggles are. We should assume the very best in everyone. We should be more like our cats—ready to engage. It is so easy to make a snap judgment of a person by the clothes they wear or the words they use. But, we are often wrong when we do this.

A broken down old man, unshaven and wrinkled from head to toe, came into the clinic one day with a soiled and battered cardboard box. In the box was a dead cat. The man was crying and he said he'd just lost his best friend. He was clearly devastated by the loss and he needed to talk about his final minutes in order to have closure. It sounded to me like the cat had suffered a cardiac event and had died quickly with no warning signs. We talked for a little bit and then he left his cat with me for private cremation. None of us expected him to return for the ashes or to pay for the cremation. We had thought he might be homeless. A few days later he returned, shaved and well dressed, and asked for his cat's ashes. He brought us his cat's carrier and some flowers and thanked us for our help. I was very touched and I was also a little ashamed because I had made a judgment. I don't know how many times I made similar judgments by assuming that someone could not afford or would not pay for recommended treatments. I was more often wrong than I was right about that. It was more often that people of some affluence chose not to treat their pets based on costs or left without paying. But, again I judge. Money makes everyone crazy.

The final lesson from the cat is that we should purr, not hiss. This is slightly different than his instruction to sing. Cats try very hard to get along. Yes, they have strict codes of conduct. They don't suggest you need to be friends with everyone. But, they are reasonable. They are tolerant. They are peacemakers and peacekeepers. But they are not pushovers.

Sometimes they know that it is best to walk away and to agree to disagree. Aggression comes in so many forms just as we find in cats out of synch with their true natures or in those who have had poor beginnings. It may feel like there is little we can do to stop it. But what about the lady who steals a parking spot, the telephone marketers, the driver that cuts off our view at a turn, the dog that steals the birthday cake (a cat would never do that), or the T.G.I.F. client that so many of us are familiar with in our various workplaces and personal lives? These petty annoyances and hostilities show us just how patient and compassionate we really are, and that they are perfect opportunities for growing compassion and peace. Perhaps individuals who challenge us are doing their best in that moment as well. They are our next lesson and our challenge—not our next annoyance, and we should respond instead with grace. What if we smiled? That would feel much better for each of us. It takes a few more minutes to be patient and thoughtful instead of being irritated or aggressive, but there is time for it. And it would be a much nicer habit? Our cats do this. They assume the best. They work with what they've got. They often pick the person who likes them least in fact—the person who is disinclined to notice or appreciate his efforts—and he works slowly on them and softens them like putty in their paws.

I am sure the cat finds it tiresome to try his best to be loving and patient and to still look around him and see his humans continue to respond in ways that make him cringe. It must be tempting to just throw in the towel and become the grump he knows he can be, but the cat never gives up. He tells us to think of the small sphere. He says we can make a difference and that we shouldn't give up. Or maybe he is saying to those of us on the verge of giving up, "Perhaps they need a cat." We are bound to mess up sometimes, because we are only human after all.

Mother Teresa said, "If you are kind, people may accuse you of ulterior motives, be kind anyway. If you are honest, people will cheat you. Be honest anyway. If you find happiness, people will be jealous. Be happy anyway. The good you do today may be forgotten tomorrow. Do good. Give the world the best you have and it may never be enough. Give your best anyway. For you see, in the end, it is between you and God. It was never between you and them anyway."

Chapter 11

If I did not have my cats to remind me of these simple truths, I would read this quotation every day. It is tempting to expand this idea of cat peace into all forms of violence—even those we perpetrate against ourselves, our lands, air, waters, and the animals we share this world with. But the cat reminds me to be subtle so I will just say that I wish we could go back a few decades and come forward with the footprint of a cat. That would also feel good. Knowing what we know will hopefully allow us to proceed more cat-like in the future. Think of the small sphere of influence and the idea that we must first do no harm wherever we find ourselves. If we can do that, will it not make a difference in the large sphere? Yes, it will, because spheres overlap.

We have heard the teachings of the cat over and over again in countless different ways and from countless directions. They remind me of certain parts of the Bible for instance—not the 10 Commandments exactly—they are softer and subtler suggestions rather than commandments. This does not surprise me. Cats are divine creatures. God works through them like he does through us. But then, we can find divinity wherever we look for it. These lessons are a collection gathered again and again from the many cats that have come into my life over the years; they each live their nine lives and give the same lessons. They are the wisdom of the feline. They may be the secret variables to stretching out that finite number of heartbeats.

I will not say that I missed my calling or that I was meant to do some other thing. I do not think my virology professor's prophecy was entirely right. Veterinary medicine has been a beautiful career for me for many, many years. But, I have finally given myself permission to own the particular weakness that first called me to the field, and now has called me away from veterinary medicine, and hope that it will be my greatest strength in writing and other endeavours. The cat has taught me that when we stray from our true natures, that is when we become imbalanced.

In many ways, the cat saved me. He taught me his language. He helped me find my words, and now the only danger is that I may never shut up. It is amazing to me how when we are on track, things just flow and come easily, and when we are not, there are obstacles that come and go, which impedes us. Once I decided to let my clinic go, it took five minutes—a single phone call—to find its new most wonderful owner. This was the law of attraction at work again. Trust in your instincts, like a cat does.

The Energy of a Cat

There is no end to the combinations and permutations of feline personalities and energy, their coats, their faces and musculature, their tails and whiskers, their gaits, and their meows and purrs; they are all so magnificent and awe inspiring. Like snowflakes, no two are alike. We are also like that. The look of a cat strutting nonchalantly towards me always takes my breath away; that tamed wildness.

Just as beautiful are the inner complex workings of the body, a perfect circuitry of life that never ceased to amaze me. All that beauty and elegance cannot possibly end with the last heartbeat or brainwave. You know those waves; you've seen them on EKG and EEG machines that are hooked up to patients on popular TV dramas. Those measurable waves of energy that form our second self, our true self, and our energetic self; they go on and on once they are shed by our earthly form. We emit this energy all the time, not just when machines measure it. Our energy field stretches out at least nine feet beyond our physical self. Think of that. Although I warn you, once you start thinking about it, it is difficult to stop. But that is a good thing.

The first law of thermodynamics states that energy can neither be created nor destroyed; it is conserved. It does not end. Those waves, after all, are just invisible atomic particles that connect briefly with other atoms adjacent like an invisible game of leap frog or the contagious nature of a smile passed on to the next across a room and beyond.

The contagious nature of energy can be observed by watching cats work their magic on humans. I have seen it. The cat's irresistible positivity and unconditional love can purr a timid and fearful child into a confident fearless adult. He can coax a grumpy reclusive old man into being a socialite again who seeks family and friends who also need him in their lives. He can spin a lonely depressed, disappointed divorcee into someone's beloved fiancé. I have seen cats work their wonders hundreds of times. And I don't even know all their stories. Humans don't always share. But I saw many stories of redemption and transformation play out before me, and it started with a cat. It doesn't have to be a cat, but this is the vehicle of energy I studied and the cat was my first teacher.

Energy is like that, and like our cats, we are energetic beings. We can be spheres of positive energy. If we could reduce life down to the lowest common denominator, it would be those particles that create energy waves like we see on an EEG or EKG. We are all energy emitters and receivers. The energy that we choose to emit every day is who we are. It may not be who we think we are, or who we plan to be, but it is who we are. The cat's energy in his natural state is always one of grace, beauty, optimism, and joy. The fact that they are sometimes aloof only increases their appeal and the intrigue we have towards them.

Originally, it was the cats' ability to keep rodents under control that raised them up in human society. But it is these other magnetic qualities and their radiance that have ensured that the cats continued to have popularity and success as a species. Many feel that cats are not fully domesticated and could break away at any time and fend for themselves, but this is flawed thinking. Even feral cats live as close to humans as they feel safe in order to profit from some human connection. They need us. We benefit from the connection as well. Kindhearted self-appointed caregivers to feral groups fully understand the feral cat's plight and feel frustrated by the general lack of awareness to it. They see its genesis. They see neighbourhood cats, loosely owned, unsprayed, and unneutered running wild and breeding litter after litter. They see the wildness begin in litters left to their own devices with fear of humans beginning at about five weeks of age if there is little or no socialization. But they also see that they are not lost causes and can be rehabilitated. Population control is a kindness and a necessity. As humans, we pride ourselves on being compassionate, but as Gandhi says, "The greatness of a nation and its moral progress can be judged by the way its animals are treated." We still have some distance to go.

Love and admiration of these mysterious other-worldly creatures began with the ancient Egyptians. In that culture, cats were so well loved that they were worshipped as god-like creatures. Killing a cat was punishable by law and warranted the death penalty. The secret language of the cat was sacred and so well protected that it was once illegal to export cats out of Egypt. The popularity of the cat has risen and fallen over the centuries, but they have persevered and once again enjoy huge popularity worldwide. They still elicit reverence and respect in anyone who learns their language. No longer expected or even encouraged to hunt and control rodent populations, the modern housecat now enjoys a prestigious figure head role within our families—that of the companion. The cat has adapted itself perfectly to this pampered lifestyle. You could say that they are the ultimate success story. We cater to their every need. It is often said that dogs have owners, and that cats have staff. However, if the environment is not exactly in harmony with their natural independence and prevents them from fully expressing themselves, then they may become ill.

Observing cats in the great outdoors and on the other side of the door can teach us about lives well lived. Perhaps that is their true purpose. I have not learned a single thing through years of yoga that I did not first learn from a cat. Both languages teach us how to tune into our intuitive intelligence, and to live well and in harmony with each other and with Mother Nature. The teachings of yoga and the cat remind us we are all a part of Mother Nature, like one huge organism. Each of us brings something essential with our gifts and paths to make every system within it work like the inner workings of a clock; our own bodies and beating hearts are just a miniature version of that bigger world. But we have forgotten that we also need to interact with nature for our own systems—our own clocks to keep working properly and to have balance. There is a necessary symbiosis there that requires our feet to sometimes get dirty. Both languages, the cat and the yogi, are so powerful and sacred, and were once secret.

The cat is also fully fluent in the secret language of disease. This means that the cat shows us, in this sign language by his own health issues and behaviours, that his modern lifestyle can create or at least contribute to disease. The cat is an interpreter of this hidden language of disease. After so many years of working in a feline only practice I certainly developed bias regarding indoor versus outdoor issues. It was impossible for me not to draw parallels with our own often deficient modern human lifestyle and health concerns. The cat shows us in his behaviors and his own body in disease to look at where we might also be going wrong ourselves in our modern day lifestyle and environment. He shows us that this may be causing similar, often vague, or poorly understood but potentially serious disease processes, like inflammation, within ourselves. The cat's life indoors is less complex and varied than our own lives and therefore is a more honest account of risk.

Cats are thought to have nine lives because they are rigorous. They do become ill from time to time with various ailments often due to imbalance as we have seen, but with care, they can do well again. They are good healers. I like the idea that there are many chances to do well or to succeed. This is true for humans as well. Taking chances and living fully with intent and persevering in the face of small and great obstacles can reap great rewards. I think of the cat as the ultimate inspirational or motivational speaker. We just need to first understand his subtle language in order to benefit from what he says.

If we imitate the cat and become a copycat by radiating a loving and open kind of energy even to those we do not know or to those who challenge us, then something beautiful happens. It is reflected back more times than not. A smile is a good example. A purr is another. And so this will advance our own healing of whatever ails us as well as those around us. Our sphere of influence may be small but as spheres overlap, there is a domino effect that spreads positive energy in every direction. And this is why I feel every household could benefit from a cat.

We feel the loving healing energy of the cat. We can pass it on—the gift of subtle positive energy and compassion in small kindnesses—like our cats offer to us every single day. This is my new recitation. It is the one that I chant humbly, silently, and without a stage. It replaces the many well-practiced recitations given again and again in my exam room.

I am grateful for having had the unique opportunity to study life up close and in an accelerated fashion. Life goes by so quickly when you are a cat. Those curious mischievous kitten days with all their challenges and lessons are quickly followed by the more experienced and thoughtful, but they still grin-eliciting adult days. In turn, these days become the wise graceful senior days of the Guru Cat. Our nimble felines blend effortlessly into our lives and they gently nudge us to be our full selves as they are, whatever our situation or circumstances. I feel I have lived through thousands of life times through the lives of my patients. I have a heightened awareness of what it means to live and what it means to die, and the mechanics of it all, but also the preciousness of it, each breath, each heartbeat, each minute or hour of wakefulness and I am fully awake and present. Life is a gift not to be wasted or put away in a drawer barely lifted from the wrappings within its beautiful gift bag. Let it out of the bag. We should use all our nine lives as our cats do and be fully worn out by the end of our last heartbeat. If we measured our age by those indices—the heartbeat and the breath—instead of years, then we would learn a little earlier to have gratitude for each new day. We would use this gift of life better and to more purpose. We would not waste our breath. This is the secret language of the cat. Pass it on.

References and Further Reading

Books:

Beaver, Bonnie V. Feline Behavior: A Guide for Veterinarians. Second Edition. Missouri: Saunders Elsevier Science., 2003.

Goldschneider, Gary and Elffers, Joost. The Secret Language of Birthdays. New York: Viking Penguin Books, 1994.

Lappin, Michael R., DVM, PhD. Feline Internal Medicine Secrets. Hanley and Belfus, Inc., 2001.

Ober, Clinton, Sinatra, Stephen T., and Zucker, Martin. Earthing: The most important health discovery ever? California: Basic Health Productions Inc., 2010.

Somerville, Louisa. The Ultimate Guide to Cat Breeds. Regency House Publishing Ltd., 2002.

Websites:

Careforcats.ca

Feline Advisory Board website: www.fabcats.org

Indoor Pet Initiative website: indoorpet.osu.edu/cats

Author Biography:

Dr. Carol Teed graduated from the Atlantic Veterinary College in 1990. She spent most of her twenty-three year career as a veterinarian working in a feline specialty practice where she became fluent in the mysterious language of cats. However, she maintains there is no such thing as a feline expert and refuses to call herself one even after seventeen years owning and operating her own feline practice. While observing cats through their accelerated life cycle, she became fascinated by their ability to bring positive change to the lives of those they interact with. She currently lives in the rural Niagara region, in Ontario, Canada with her husband, four children, two dogs, three cats and three chickens. She is working on her second book, and always has a cat on her lap and two looking over her shoulder.

If you want to get on the path to be a published author by
Influence Publishing please go to
www.InspireABook.com

Inspiring books that influence change

More information on our other titles and how to submit
your own proposal can be found at
www.InfluencePublishing.com

CPSIA information can be obtained at www.ICGtesting.com
Printed in the USA
LVOW06s0137201113

361802LV00008B/36/P